Grand Army of the Republic

AND

Union Veteran Research

The Essential Guide for Genealogists and Family Historians

Gary W. Clark

PhotoTree.com™

Grand Army of the Republic and Union Veteran Research
The Essential Guide For Genealogists and Family Historians

Library of Congress Control Number: 2021914078

www.PhotoTree.com

Published by PhotoTree.com

ISBN: 978-0-9835785-4-3

Cover and interior design by Gary W. Clark.
Edited by Gena Philibert-Ortega.

Cover Photos:	Grand Army of the Republic membership badge. Woman's Relief Corps membership badge. Ladies of the G.A.R. membership badge. Badges from Gary W. Clark collection.
Back Photo:	Stephenson Grand Army of the Republic Memorial, Washington, D.C. (Library of Congress).

Acknowledgements

I wish to thank my editor and genealogy mentor Gena Philibert-Ortega for her valuable input and guidance which kept me from wandering too far from my actual goals. Also, several other colleagues have contributed their expertise and keen eyes to the topics and book. Fellow genealogist Margaret Kline offered her expertise in military research to ensure completeness and accuracy in many areas. My friend in genealogy and research adventures, Anna Uhlig, applied her sharp eye for clarity. And a distinct appreciation goes to David Allen Lambert, Chief Genealogist of American Ancestors by New England Historic Genealogical Society, for his review of the book. His extensive knowledge of military research is greatly respected in the industry and invaluable.

I have benefited from numerous organizations, universities, agencies, and libraries for assistance in research and acquiring images. The genealogy department at the Wichita Public Library in Wichita, Kansas was exceptionally helpful in searching for photographs and books in their archival vault. Access to many Library of Congress records and images was important to the content of this book. Also, several internet sources for archived articles, stories, and information were indispensable. These included Internet Archive, HathiTrust Digital Library, and Newspapers.com.

Particular thanks goes to the staff of the Harvey County Historical Museum, Newton, KS, for granting me unlimited research time into a phenomenal collection of original G.A.R. and W.R.C. records and journals. The staff of the Kansas State Historical Society, Topeka, KS, was also very gracious during the pandemic lock down in 2020 for allowing access to original documents, certificates, records, and journals in their Grand Army of the Republic collection.

Gary W. Clark

Contents

If you are researching a Civil War veteran you need to understand the G.A.R.

I Introduction

The *Grand Army of the Republic* was a major force in the lives of Civil War veterans for more than 50 years, and its influence on the whole of America was equally large. It has been credited with the election of at least four U.S. Presidents, was the champion for Memorial Day becoming an American tradition, became the center of politics in many towns overshadowing actual political parties, organized some of the biggest gatherings across the country, and became an undeniable force honoring Union soldiers. Its presence is still evident in cemeteries and town squares across the country.

G.A.R. Postcard – Postmarked May 30, 1910, Memorial Day

Like so many wars, the average age of Civil War soldiers and sailors was about 20 years, hence many veterans lived well into the 20th century. One of this author's great-great-grandfathers, a soldier in the 56th Illinois Infantry, lived until 1941 at 95 years old.

The Grand Army of the Republic (G.A.R.) represented many of those veterans, however the number of members over its nearly 80 years of existence is not easily calculated. The G.A.R. itself claimed over 400,00 members in the early 1890s, but this was simply a snapshot in time. Many members, over the course of several decades, joined early, dropped off, and then joined again. Probably more than half of all Union veterans at one time or another belonged to the Grand Army—leaving an amazing yet jumbled trail of records.

If you have an ancestor or research subject that served in the Union Army or Navy during the Civil War, there is a good chance they were a member of the Grand Army after the war ended.

By the time the confederacy surrendered, over 1,600,000 Union soldiers and sailors had survived the war and were soon to become veterans. As with any cessation of war, there was now the task of putting a normal life back together for these men. Unlike American wars that followed, the Civil War was waged on its own soil. Reminders of the war were always around, even if just in a soldier's memories. Yet, there was no support system to help them transition back to pre-war civilian life other than pensions for those with permanent disabilities. It was up to individual veterans to heal and thrive in a now-different environment.

The desire to gather with a welcoming group of like-minded veterans brought many veterans back together. Soon, the Grand Army of the Republic became a source of camaraderie and resounding voice of veterans. Indeed, fellow G.A.R. members were referred to as Comrades.

G.A.R. Membership Badge

Genealogists and family historians benefit from millions of G.A.R. documents that detail where the members enlisted and lived, many of their activities, and histories of their service.

Many researchers, especially those who have not served in the military, do not realize the magnitude of military record keeping, even back hundreds of years. There is a timeworn adage that an army travels on its stomach, referencing the fact that soldiers must eat. Every C-Ration, can of beans, and even the bullet that killed the deer for a meal, were all available only because someone filled out a requisition form, a shipping document, and a distribution order. Soldiers and sailors hardly took a step without a trail of orders directing them to do so. Afterwards, a report was made about how well that order was executed.

Of course, the G.A.R. was organized by ex-soldiers for ex-soldiers. Old habits did not die after the war; the reports kept piling up. G.A.R. Muster Rolls, an attendance report of sorts, may tell you where your great-grandfather was every 3rd Saturday night in 1889. Any good genealogist will tell you that discovering an original source document such as a Muster Roll is like finding historical gold.

G.A.R. records usually show when and where they joined, when they left, and when members died. Also, their occupation and where they lived was recorded. Other entries reveal the military unit they served in, wounds and injuries suffered, dues they paid, and other noteworthy details. In many cases, these records are like a diary of a veteran's post-civil war activities.

The U.S. government, mostly through the National Archives and Records Administration (NARA), has assembled and organized a vast collection of veteran military records. To help with your research, the steps to finding a veteran's Compiled Military Service Record (CMSR) and pension record are included in this guide.

Yet, there is no single comprehensive body of G.A.R. records since the organization was not part of the federal government.

G.A.R. archives, some from small rural posts, are scattered across the country in various places, including forgotten records awaiting rediscovery in court houses, small local libraries, and other places near the original source. Still, others have been meticulously cataloged and organized for easy access by state archives, libraries, historical societies, and online companies that provide genealogy and family history data. However, finding all these resources is difficult for the average researcher as most are compiled state-by-state, sometimes county-by-county. The devil is in the detail of finding local records. To help the researcher, Chapter Eight is a state-by-state guide to the most informative sources.

A thorough story of the G.A.R. cannot ignore the wives, parents, children, and fellow Americans who wished to acknowledge and contribute to the welfare and memories of veterans. Affiliate groups such as the Women's Relief Corps (W.R.C.), Ladies of the Grand Army of the Republic (LGAR), the Sons of Union Civil War Veterans (SUVCW), and the Daughters of Union Veterans of the Civil War (DUVCW) worked alongside the G.A.R. and each continue in some form today. Their records may include the names and deeds of those women and men which add even more substance to your historical narrative.

W.R.C. Membership Badge

After years of chasing ancestors, exploring cemeteries, visiting archives, and helping others in their quests to assemble an ancestor's history, I decided to document my experiences and knowledge so that a wider group of people could benefit.

This book brings you the history and record sources of the G.A.R. and its auxiliary organizations. However, it was difficult to solely focus on the Grand Army of the Republic and its members, and not explore the greater collection of Union veteran records.. While I considered the G.A.R. records an untapped source of information, the most common Civil War records, usually held by the National Archives, needed to be addressed also; giving the researcher one place to begin and complete their investigative journey.

Any genealogist, family historian, or history buff will benefit from this time saving and comprehensive guidance to finding valuable and sometimes well-hidden information about a Union veteran. I wish you the elation of success that I have enjoyed.

Gary W. Clark

Descendant of two G.A.R. members.

Dedicated 1913, *"In Memory of the Union Soldiers and Sailors of the Civil War."* This monument was erected in Wichita, Kansas for $20,000. Inscriptions include *"Grand Army of the Republic ~ Fraternity Charity Loyalty"*. The familiar G.A.R. medallion is embossed at a height of three feet on the side of the base. (Gary W. Clark collection)

1 A Brief History of the G.A.R.

A Soldier's Vision • Political Redirection • Near Collapse • Pensions to the Rescue • An Inevitable End

Little time passed after Lee's surrender at Appomattox on April 9, 1865 until Union veterans reclaimed the bond of camaraderie they experienced during the war.

Post-war America started with a huge parade on May 23, 1865, labeled *The Grand Review*. Over 150,000 Union troops rode and marched down Pennsylvania Avenue to throngs of cheering citizens, expressing gratitude to those who served and sacrificed. This set the stage from which veteran organizations could spring.

The Grand Review – Washington D.C. May 23, 1865 (Library of Congress)

Whether they were looking for friendship with those of common experiences, or a union to gain service benefits, or an avenue to solicit recognition, there was a natural magnetism among the veterans. This phenomenon is not limited to military-related groups. School alumni, business associations, hobby clubs, and of course well-known fraternal organizations embody the same inclination. Social organizations become a second family for many members.

A unique quality of the Grand Army of the Republic (G.A.R.) is that it survived organizational dysfunction, political controversies, and financial disasters to last over seven decades. This could not be said for countless other groups formed after the Civil War.

The G.A.R. was not the first society organized by Civil War veterans, not even close. Scores of local, regional, and a few national organizations cropped up, a few even before the war ended. Some were associations of specific military units or commands, others had geographic or professional ties.

- Third Army Corps Union ,1862
- Army of the Potomac, 1862
- Society of the Army of the Tennessee, 1865
- Military Order of the Loyal Legion, 1865
- Soldiers and Sailors Republican Convention, 1866
- Grand Army of the Republic, 1866
- National Association of Naval Veterans, 1866
- Society of the Army of the Cumberland, 1868
- Society of the Army of the James, 1868
- Society of the Burnside Expedition, 1869
- Society of the Army of the Potomac, 1869
- Society of the Army and Navy of the Gulf, 1869

To receive a G.A.R. charter, local groups of veterans were required to assemble only ten veterans. Like many organizations, the Grand Army's drop-off rate was high no matter how dedicated the founders were in celebrating their cause. Many veterans simply lost interest, moved west or died. However, with well over one million veterans spread across the country, the opportunity for growth outweighed turnover. The Grand Army offered a scalable organization that veterans could understand. It addressed their local needs, yet extended its influence into the highest office in the land.

G.A.R. membership amounts to an extension of the military career of a soldier . . .[1]

A Soldier's Vision

Two members of the 14th Illinois Infantry, Major B. F. Stephenson and Chaplain W. J. Rutledge, were part of General Sherman's *Meridian Campaign* (Mississippi) in February of 1864; a prelude to Sherman's *March to the Sea.* Legend has it, as tent mates during a lull in the action, their thoughts turned to life after the war that included a post-war organization of veterans.

Major B. F. Stephenson and Chaplain W. J. Rutledge[2]

> *Chaplain Rutledge suggested that when the troops were finally mustered out of the service it would be natural to suppose that men who had shared so much suffering, privation, and danger would wish to form some sort of association, that they might meet again to preserve the friendships and memories of the past.*[3]

Indeed fraternity, a feeling of friendship, became the first tenet in their *Three Guiding Principles: Fraternity, Charity, and Loyalty* or FCL. This monogram was emblazoned on badges, monuments, building, and printed on countless documents.

Stephenson Grand Army of the Republic Memorial, Washington D.C.
(Library of Congress)

Discussion continued of their idealized fraternity during the war and with later correspondence. After their discharges, Reverend Rutledge returned to Petersburg, Illinois, approximately 25 miles from Major Stephenson's home in Springfield. They reunited nearly a year after the war's end to work on specifics of their dream. The Major, a medical doctor, had already become the primary force behind the plan and engaged many people in the Springfield area and surrounding states. Detailed versions of the organization's mission, structure, and even initiation rituals were already in draft form.[4]

The initiation rituals and indeed the *Ritual* book were nearly complete and soon to be published in Decatur, Illinois by a veteran-owned printing company under strict secrecy. Shortly, on April 6, 1866, the first Grand Army of the Republic post was chartered in Decatur. This post was actually in existence while the Grand Army of the Republic's Constitution was still being revised as it wrestled with organizational and reporting structures, goals and political involvement, and the need to create a sustainable membership. While it was necessary to adjust goals and tactics, the constant revision process would plague the G.A.R. throughout its early existence.

The basic G.A.R. reporting structure change little over time. It was a typical hierarchy of command and reporting from National Headquarters to local units.

- National Headquarters – In the early years it was located in the National Commander's hometown.
- Department (State) Headquarters – This was the state organization, sometimes two states with few members were combined into one Department.
- District Headquarters – County headquarters, soon eliminated as too bureaucratic.
- Post – A local unit chartered by a minimum of ten veterans.

The Grand Army went through many cycles of trying to be a veteran's benevolent club, a lobbyist for veteran benefits, a cloaked political party, and even a battle-ready militia group. These winds of change usually came from the desires and goals of top leadership who seemed to easily 'sway the troops' under the guise of being the country's saviors.

But one constant remained throughout the G.A.R. life—a tremendous number of records were produced. The insatiable appetite for records and reports would be an *Achilles heel* for many posts as we shall see. Yet those same reports are what genealogists and historians seek.

G.A.R. Membership Eligibility

The Grand Army of the Republic had two simple eligibility rules:

1. Honorably discharged veteran who served during the Civil War.
2. Had never borne arms against the U.S., at anytime.

The exact regulation is more detailed; the following is from Chapter I, Article IV from the 1884 edition of *Rules and Regulations*, commonly known as the *Blue Book*.

> *Soldiers and sailors of the United States Army, Navy or Marine Corps, who served between April 12th, 1861, and April 9th, 1865 in the war for the suppression of the Rebellion, and those having been honorably discharged therefrom after such service, and of such State regiments as were called into active service and subject to the orders of United States general officers, between the dates mentioned, shall be eligible to membership in the Grand Army of the Republic.*
>
> *No person shall be eligible to membership who has at any time borne arms against the United States.*[5]

Clarification and Judgments

Like most regulations, clarification or rendered judgments were needed as objections or exceptions were filed. The G.A.R. answered these with official *Opinions* from the Adjutant General, usually published during a National Encampment.

Ex-Confederate Soldiers

An early eligibility challenge was of soldiers who served the Confederacy, but changed sides and served the Union. The requirement that no person "has at any time borne arms against the United States" was strictly enforced. Ex-Confederates were not eligible.

Independent Fighters

Other classes of fighters were decided upon according to the circumstances and set of facts. Those included local militias, scouts, privateers, home guards, and other forces. Generally, a soldier needed to be part of a unit that reported to the Union Army for command and deployment, and was paid by the Union Army.

Women in the G.A.R.

One of the popularly followed questions was whether women were eligible for admittance to the G.A.R. There was not a specific rule that excluded women from the G.A.R., but they were essentially refused entry because they could not have been legitimate, honorably discharged veterans of the Union Army. The Civil War-era Army did not allow women soldiers. Therefore, none were eligible to join the G.A.R.

Now for reality. Reportedly, hundreds of women fought for the Union Army, under many different circumstances. Some fought along with their husbands, others were unofficially allowed in close knit companies, or were associated with a regiment but participated in fighting at some critical juncture. However, they were not inducted or enlisted soldiers, and no discharge was granted, which was an eligibility requirement.

Daughter of the Regiment

There was a practice in many regiments of wives, daughters, or other women traveling with a regiment, often providing cooking services, medical help, or other unofficial support roles. They were kindly called *Daughters of the Regiment.*

Opinion XXXI, issued April 15, 1872 by the Adjutant General stated:

> *I do not know of any position in the army, according to the Regulations, described as daughter of the regiment. Our Regulations prescribe as a condition for membership that the applicant shall have been a soldier or sailor in the army, navy or marine corps of the United States, etc. Whatever a daughter of the regiment may have been, I do not understand that she was a soldier in the United States army. She may have done good service in the suppression of the Rebellion, as nurse in hospitals and agent of the Sanitary Commission and Christian Commission, and many patriotic citizens at home did; but as she does not come in the classes of persons whom our regulations make eligible to membership . . .*[6]

Disguised as Men

There appears to be one verified account of a woman in the G.A.R. Sarah E. Edmonds joined the 2nd Michigan Infantry under the guise of a man, using the name Franklin Thompson. She served two years and participated in numerous campaigns. Her story is long; she even wrote a book afterwards titled *Nurse and Spy.*

Sarah is listed in the National Park Service's Soldiers and Sailors Database and the Compiled Military Service Records (CMSR) in the National Archives, where Franklin Thompson is documented with the 2nd Michigan Infantry.

Many newspapers covered her story and death in 1898, and noted she was a member of the George B. McClellan G.A.R. Post of Houston, Texas:

> *Her funeral was held Monday evening with military honors by a delegation from Houston of the Grand Army of the Republic post to which she belonged.*[7]

There are other cases of women in the Army; a well-publicized example is Kady Brownell who along with her husband Robert, served in the 1st Rhode Island Infantry. Her situation is not clear as she was reported as a Daughter of the Regiment. Some posts may have included women as *honorary* members, although the national organization was resolute there was no such designation. Another well-known case is Albert Cashier (Jennie Hodgers) who joined the 95th Illinois Infantry as a man. She lived much of her life as a man, and was not discovered as a female until near death in her 70s while living in a Soldiers and Sailors home.

Post Admittance

Once an applicant was able to verify his service record and eligibility, he still had to be voted into a post with very few dissenting votes: a maximum of one out twenty.[8] This process quietly allowed personal prejudices to creep in which resulted in many posts maintaining a narrow character, though the G.A.R. adamantly proclaimed to harbor no discrimination of race, creed, color, or religion. Whether a restrictive practice was painfully obvious, or just subtly unwelcoming, many posts were subsequently formed with all-Black members or of a particular ethnic, national, or cultural origin.

Political Redirection

> *We are enjoying, in a land we helped to save, a companionship made sacred by common sufferings and sacrifices.*[9]
>
> Lt. Col. Robert Burns Beath
> National Commander 1883-1884

G.A.R. leaders were nearly always experts at crafting patriotic messages to rally the troops. After all, most Grand Army leaders were previously officers in the Union Army, many of them colonels and generals. Their motivational speeches to junior officers and enlisted men before a battle included the unspoken truth that they might soon lose their lives. With less on the line after the war, surely, a post-war commander could convince G.A.R. members to follow his philosophy, push for certain legislation and even vote for someone who might show favor to them.

Due to a reduction in U.S. Army size after the war, many high-ranking officers had no place to exercise the power they became addicted to. Business and industry opportunities were not as broad as today and could not absorb the large number of newly minted executives. Where else might a decorated general seek a position he felt was deserved? Politics, of course.

The war was still fresh on everyone's mind in 1868, including the actions of heroes. Moreover, the after-war chaos needed a strong leader; by default that was General Grant. That year's election hoisted General Ulysses S. Grant to the presidency of the United States, winning the electoral vote by a landslide. This theatrical poster, beautifully drawn by Charles H. Crosby, delivered a well-crafted message requiring no words to make its point.

1868 Election Poster, Library of Congress

Grant easily won reelection in 1872 against Horace Greeley, who not only lost the election, but his life. Greeley, the *New York Tribune* publisher, and subsequent congressman is most famous for promoting the quote "*Go West, Young Man,*" though he probably did not coin it. Not a young man for the time, he was 61 years old during the election. His wife died a week before the election; he became ill in despair, and saw a significant popular vote loss. While no surprise was to be expected in the results, the electoral votes were yet to be counted—he died before those results were in. One could say he did not lose the official election in his lifetime, yet he did not win it either.

While General Grant was a solid choice for the presidency, there was still consternation among many citizens. Noted author Nathaniel Hawthorne observed on a visit to Washington D.C. in 1862 that he:

> *". . . sensed an undercurrent of feverish intrigue created by politically ambitious Union officers maneuvering for military advancement."*
>
> *and visualized ". . . a government dominated for a half-century by 'an incalculable preponderance of military titles and pretensions', with a succession of 'bullet-headed' generals holding the presidency."* [10]

Strong words even for a best-selling novelist.

Since there was no war-like enemy for ex-soldiers to focus their common efforts on, the goals of national leadership and the farmer-veteran a thousand miles away were frequently not in sync. Many newly-minted politicians were simply interested in maintaining power and position, taking up issues that would most likely appeal to the greatest number of voters. Many ambitious men saw G.A.R. leadership as a stepping stone to influential and powerful elected offices.

This focus on ambition resulted in a struggle to build a unified message and organization during a formative period.

Near Collapse

While the early history of the G.A.R. may not seem useful to many genealogists, it does help explain why there are few individual records through the 1870s. Chaos did not lend itself to good archives, or to increasing membership.

The 1870s almost saw the complete collapse of the fledgling G.A.R. Different leaders came through the hall doors with their visions, with most implementations failing miserably. Economic hardships from the financial *Panic of 1873* led to the *Long Depression* that left many veterans so broke they could not afford annual dues and post taxes. The migration westward created temporary membership in many posts as settlers moved numerous times, always seeking a better farm or way of life. Also, finishing a sod dugout and feeding the family took precedence over joining a bunch of army buddies for meetings to hash over old times.

Even in those areas where the G.A.R. was taking hold, mainly in Eastern and Midwest states, there were self-inflicted impediments to smooth growth. One of the early problems in the G.A.R. was the implementation of a system of *Ranks* in 1869 that had little to do with ranks that veterans held in the service. These ranks were borrowed from fraternal societies such as the Freemasons *Grades* system. This was meant to establish order and discipline in a G.A.R. post; the idea that it worked well for the Masons was thought to elevate the Grand Army of the Republic to a highly respected association. Yet it coincided with the beginning of a downward spiral for the G.A.R.

The Ranks system did reflect military terms, yet in some manners this made it worse. The Rank system was not overly complex, it consisted of three levels: *Recruit, Soldier,* and *Veteran.* A new member was initiated as a Recruit, which he remained for two months. Then a recommendation by two members and approved by two-thirds of the post was needed for advancement to Soldier.

Now, those with military experience know that a recruit is lowest form of life in the service—just ask any basic-training drill sergeant and they will tell you so. A battle-hardened veteran who may have bravely progressed through the Army ranks did not take kindly to being called a recruit again. Especially since this rank held no privileges or voting voice in the post.

Popularity contests and personality judgments soon crept in. If someone had a brusque or surly disposition, he may not have cultivated friends for needed votes, though he had been a good soldier in the war. If he did advance to Soldier, he then waited six more months for elevation to Veteran, again by the same procedural rules. Complicating these accrued abuses further, a newly promoted veteran was required to pay a surtax on top of their membership dues at each step. At least in the Army, new soldiers did not have to pay extra for insults, they were free. There is no way to determine how many new recruits soon left due to these insults, but the situation was not positive and is acknowledged by historians to have caused the G.A.R. to slip in membership.

Death by Bureaucracy

In addition to the insulting combat veterans to a new level of disrespect, other bureaucratic processes were not attractive to many veterans. The Masons and many other organizations were built around tradesmen and businessmen in larger cities to foster professional connections and economics. They embraced rules, secret rituals, regulations, forms, and strict record keeping. Many fraternal society members were already attuned to this type of responsibility, but many new G.A.R. posts in smaller towns on the Plains were primarily created by farmers. Record keeping on a farm was a foreign concept and certainly not an attractive one. This problem was acknowledged in the 1870 Adjutant General's report:

> *The complicated form of the reports now used by Post adjutants and quartermasters has caused great trouble, not alone to the officers making the reports, but quite as much to the officers to whom they are made, they being obliged to return them for correction, one, two, three, or four times.*
>
> *I would suggest the propriety of appointing a committee by the National Encampment, consisting of the assistant adjutant-generals who may be present, to whom all forms for Post reports be submitted, for such revision as they may deem advisable to recommend to the Encampment.*[10]

Appointing a committee to study the problem did not lead to immediate and useful reforms. Indeed, the Adjutant General's report at the 1874 National Encampment noted about Kansas:

> *Here is a Department Organization, but no reports have been made to National Headquarters for two years.*[12]

This was not an isolated case. By 1875, the departments of Delaware, Indiana, Iowa, Kansas, Maryland, Michigan, Minnesota, Missouri, and New Mexico were defunct.[13] And many other states reported fewer than ten posts each.

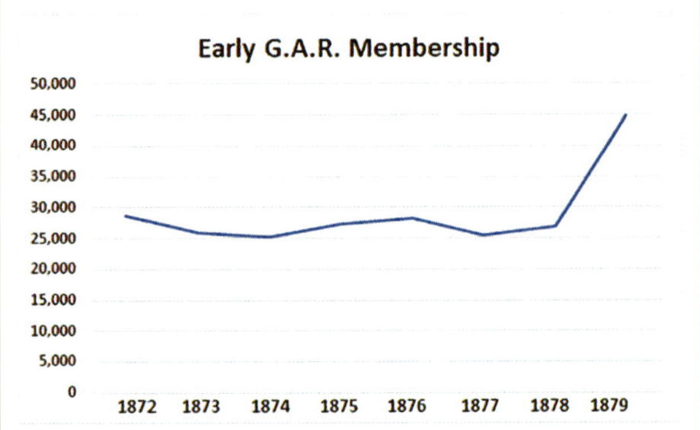

The G.A.R. did not have an accurate membership count in the early years. In 1872 they finally began reporting their total membership, culled from questionable and sometimes incomplete Department reports. With membership hovering below 30,000 before 1879, this number represented only approximately two percent of the total number of surviving veterans. (Data from National Encampment Journals)

Most disappointing to genealogists and family historians is the shoddy or lack of early post records containing member information. Seldom are good post records found until the 1880s.

The G.A.R. had created such a formal and redundant reporting structure that it overwhelmed average members across the country. Record automation did not exist as we enjoy today, since forms and reports were filled out by hand with information that was previously sent to headquarters many times over. Some Post Adjutants became creative by noting on reports "same as last quarter," but many posts simply did not send reports at all.

In addition, with the attention paid to political in-fighting by national leadership, the farmer/veteran at distant posts was not seeing any benefits, except drinking beer and swapping lies at Saturday night post meetings. There was no need to file reports for those activities.

Stickiness Prevails

Malcom Gladwell, a well-known social sciences author, refers to *stickiness* as a powerful trait that makes a movement or product a success.[14] A college alumni association succeeds because of a common bond among members that lasts forever. A Civil War soldier's military experience stuck with him, creating a bond with his G.A.R. comrades.

Chaplain Rutledge's idea "*that they might meet again to preserve the friendships and memories*" was a key component of the G.A.R. It was repeated frequently, and even became part of a recruit's initiation ceremony, conducted by the Officer of the Day with sword drawn:

> *Good men and true who having served faithfully in the Army or Navy of the Union during the dark days of the rebellion, now seek to unite with us, their late comrades in arms, in preserving the memory of those hours of trial and danger.*[15]

Preserving the memory became a core component to the long-term survival of the G.A.R.

Ex-soldiers and sailors would always be veterans. They could not resign from their experience; once a veteran, always a veteran—it stuck to him.

This unending link opened the door for a veteran to become or rejoin as a member of the Grand Army. Muster records show that many members joined, left for an extended time, then came back as members in good standing.

Pensions to the Rescue

The Grand Army became a potent political force in the latter part of the 19th century. More than any political party or special interest group, it controlled the fate of many politicians and therefore legislation. This was not by accident. As noted before, G.A.R. leadership was led by ex-Union officers, many which then became state and national politicians with the support of the *boys*. These politicians in turn supported or even created legislation that directly benefited veterans, solidifying further the politicians' place in power.

While political wrangling after the war dealt with reconstruction and racial controversies, by the late 1870s the G.A.R. deftly side-stepped direct confrontation with race, suffrage, and other controversial issues. Getting mired in emotion-charged subjects contributed to previous problems by dividing their membership. Their focus now turned to veteran pensions. As veterans aged and war-time injuries became more problematic, whether real or imagined, their interest in receiving compensation for past sacrifices and patriotism rose dramatically.

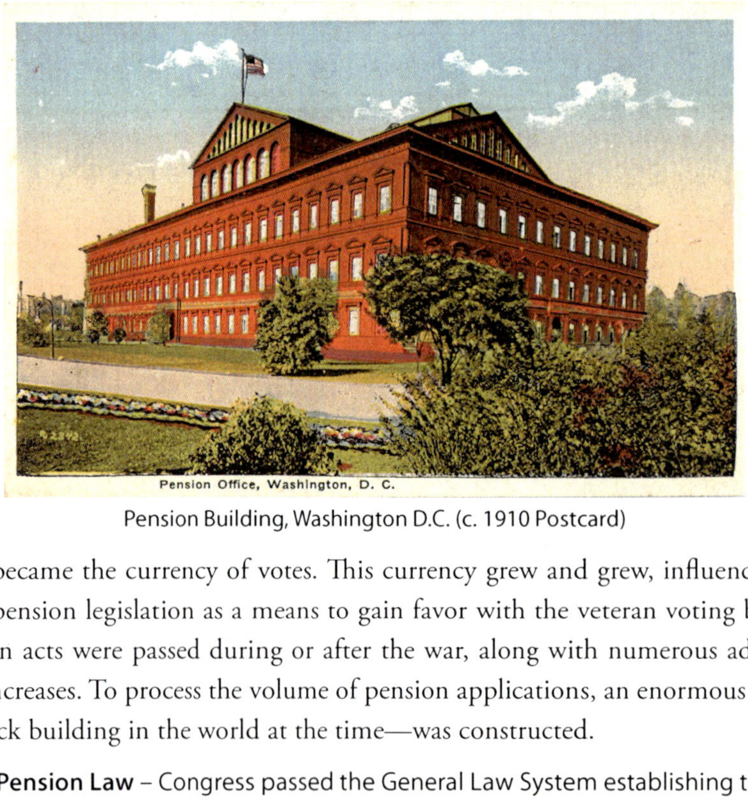

Pension Building, Washington D.C. (c. 1910 Postcard)

Pensions became the currency of votes. This currency grew and grew, influencing politicians who saw pension legislation as a means to gain favor with the veteran voting bloc. Four major pension acts were passed during or after the war, along with numerous adjustments and pension increases. To process the volume of pension applications, an enormous building—the largest brick building in the world at the time—was constructed.

- **1862 Pension Law** – Congress passed the General Law System establishing the Pension Bureau and granting pensions to soldiers with war-related disabilities and dependents of those killed on duty. Full disability garnered $8 a month, equivalent to over $200 in 2021.

Additional acts were passed after the war as aging veterans became more reliant on pensions.

- **1879 Arrears Act** – The 1879 act allowed veterans to file for disability at any time AND receive back pay from the time they left the service. This gave many veterans incentive to file and receive a handsome payment.
- **1890 Dependent Pension Act** – This allowed all veterans to apply for a disability pension regardless of when an injury occurred, whether military related or not.
- **1907 Service & Age Pension Act** – All Union veterans who reached 62 years of age were eligible for a monthly pension. The amount varied by age: sixty-two years = twelve dollars per month; seventy years = fifteen dollars; seventy-five years or over = twenty dollars.
- **1912 Increase in Payment** – Monthly payments were increased.

While the 1862 Pension Law, instigated by President Lincoln, addressed wartime disabilities and death, the 1879 Arrears Act illustrated newly found G.A.R. power. Thus began a three-decade influence on congressional action, resulting in generous pension legislation. The focus on pensions helped bring the G.A.R. out of near total collapse, as veterans could easily see the benefit of belonging to the Grand Army. Monthly post meetings became the perfect place to learn about pensions; how to apply and receive them. Membership began to skyrocket in 1880. It was a win-win situation for G.A.R. veterans and politicians who voted for pension legislation.

Confederate veterans were not covered by federal pension programs. They were left to rely on their southern state for pensions.

Not only did interest in pensions increase G.A.R. membership, pension processing left behind treasure troves of records. The Pension Bureau required countless forms, affidavits, doctor reports, proof of service, address changes, and other demands that generated a steady stream of correspondence with an applicant—all saved in a veteran's pension file. These historical records are available from the National Archives, as covered in Chapter Six.

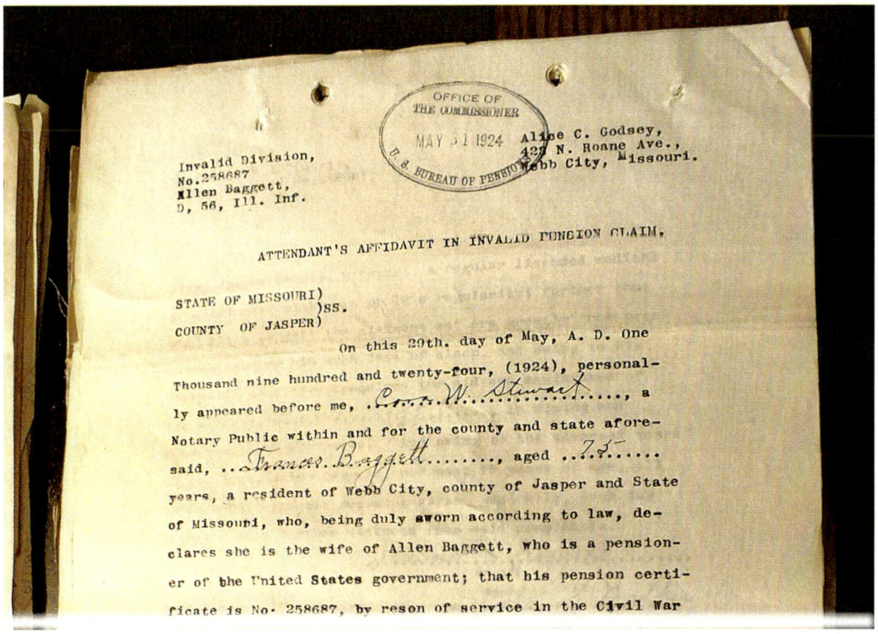

OFFICE OF THE COMMISSIONER MAY 31 1924 U. S. BUREAU OF PENSIONS

Alice C. Godsey,
423 N. Roane Ave.,
Webb City, Missouri.

Invalid Division,
No.258687
Allen Baggett,
D, 56, Ill. Inf.

ATTENDANT'S AFFIDAVIT IN INVALID PENSION CLAIM.

STATE OF MISSOURI)
)SS.
COUNTY OF JASPER)

On this 29th. day of May, A. D. One Thousand nine hundred and twenty-four, (1924), personally appeared before me, ..Cora W. Stewart......, a Notary Public within and for the county and state aforesaid, ..Frances. Baggett......., aged ..75..... years, a resident of Webb City, county of Jasper and State of Missouri, who, being duly sworn according to law, declares she is the wife of Allen Baggett, who is a pensioner of the United States government; that his pension certificate is No. 258687, by reson of service in the Civil War

A glut of correspondence was created during the pension claim process, sometimes revealing important veteran history. This author found an 'extra' wife of a great-great-grandfather in his pension file.

An Inevitable End

Few organizations define precisely when they will no longer exist. This goes against human nature. Yet, the Grand Army of the Republic had always stated it would cease when the last member died–which occurred in 1956. However, the organization had a great run and exerted a huge impact on the United Sates in a short period of time.

Once Politically-Powerful GAR Dies With Passing of Last Man

by Robert d. Price
Associated Press Staff Writer

NEW YORK, Aug 2 (AP)—The obituary notice of the Grand Army of the Republic was written today. It was born April 65, 1866 at Decatur, ILL., and died at Diluth Minn., with the passing of its last survivor, Albert Woolson.

The GAR never counted more than one-third of the Union Veterans of the Civil War on its membership roster. It is nonetheless fitting that Woolson, the last member, should also have been the last living man who wore the federal blue in 1861-65. The GAR has long since come the symbol of the Union veteran it is proper that simultaneous Taps should sound for them.

The Grand Army wasn't intended to be a political unit. Dr. B.F. Stephenson, former surgeon of Illinois Infantry who is honored as the founder, is supposed to have envisioned a great brotherhood of former soldiers, bonded together by mutual affections and memories.

The Knoxville Journal, August 3, 1956

When the last members died, post meetings faded away and national encampments ceased However, the Grand Army of the Republic left vast numbers of revered memorials that carry on its legacy. Dramatic monuments grace cemeteries across the country and stand tall in front of court houses and town squares, while countless five-pointed star markers decorate Civil War veterans' graves.

Civil War headstones and G.A.R. five-pointed star grave markers dot a large Midwest cemetery.
Photograph by Gary W. Clark

A white bronze monument erected by the local G.A.R. and W.R.C. Posts is flanked by cannons in a small-town cemetery.

Walnut Hill Cemetery, Kingman, KS (1900 Population 1,038)
Photograph by Gary W. Clark

2 G.A.R. Records

What Can You Expect To Find • The Glory Years • G.A.R. Structure • What is a Regiment • Where Are Records • National Records • Department Records • Post Records • Media Resources • Museums

Grand Army of the Republic records are one of many sets of Civil War records available to researchers. They touch a majority of Union veterans, though during the 70-plus years of the organization's existence there is no way to precisely determine the total number of members during those seven decades. Still, these records not only provide a look at many veterans, but the detail is unique since it is usually a reference to the locale where they lived and a record of their wartime experience.

Approximately 2,100,000 Union soldiers served during the Civil War; about 500,000 of them perished from battle or disease. This left about 1.6 million veterans after the war.

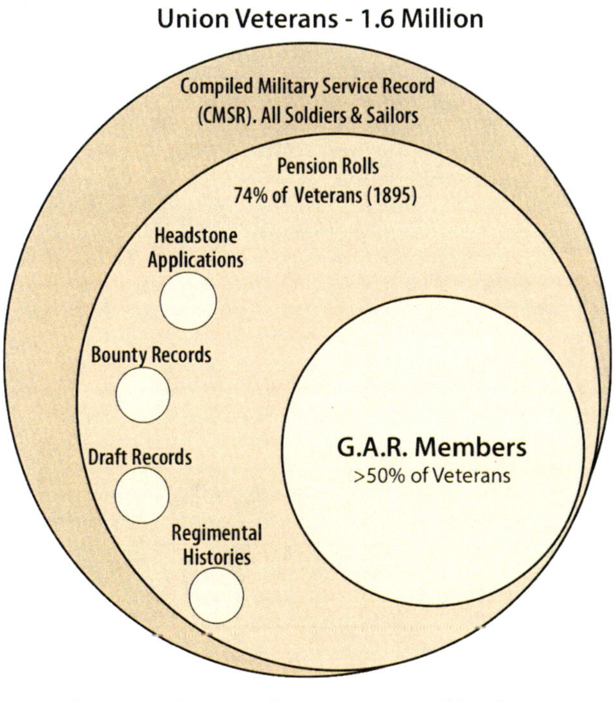

G.A.R. member records comprise one of the three largest record collections of Civil War veterans, following the CMSRs and Pension Rolls. Still, all record sets should be explored to uncover the maximum amount of ancestor history.

Since the G.A.R. was a private organization, there is no direct connection between their records and government documents that recorded their service. Yet, the conditions of membership in the G.A.R. dictated that the veteran show discharge papers or other proof of service.

If your ancestor served in the Civil War, it is very likely that a record of him can be found in one or more places, including Grand Army of the Republic sources.

G.A.R. records should be considered one of the priority search areas, along with the federal Compiled Military Service Records and the Pension Rolls. These three record groups comprise the greatest number of veterans, though admittedly the G.A.R. records are sometimes more difficult to uncover than the federal records.

Research Tip: the G.A.R. was a private organization, and not part of the Federal Government. Its archives are not among the records at National Archives and Records Administration (NARA).[1]

What Can You Expect to Find?

Part of the incentive to write this book was to help the researcher avoid futile efforts. To calibrate expectations and avoid searching for phantom records, it helps to know where the greatest number of records were created and subsequently archived.

When searching Civil War veteran information, think of your research as a series of sub-projects. Records in each sub-project are found in specific places, separate from others. Complete a sub-project with success or exhaustion, then move on to a next sub-project. To start with, one of the first pieces of the research puzzle is knowing when G.A.R. records were created.

The Glory Years

The quantity and quality of records coincided with G.A.R. growth. Therefore, it helps to understand the Grand Army lifecycle. If your ancestor died before the 1880s, he is not likely to have any G.A.R. records due to early mismanagement. From the 1870 Proceedings of the National Encampment, the Adjutant General wrote:

> *In making this report, I have found it to be absolutely impossible for me to furnish anything like a correct report of the present membership of the organization. Posts communicate with Department Headquarters and receive no answers to their communications; consequently, they feel that they are neglected, and feel further that they receive in return nothing to compensate them for their dues and reports . . .*[2]

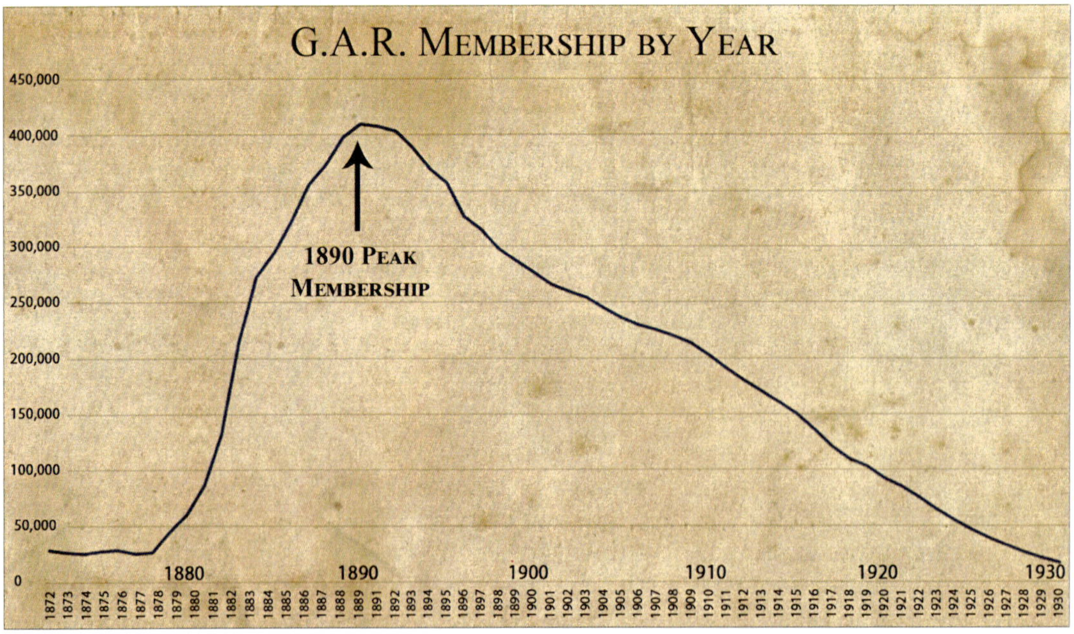

Membership in the G.A.R. by year. Compiled from 1872-1930 National Encampment Journals.

Beginning in the early 1880s, membership grew at an astounding rate to over 400,000 members, driven largely by interest in new and generous pension legislation. In addition, the record fiascoes of the 1870s disappeared as more efficient reporting processes were implemented. It is easy to see why there are more records available beginning in the 1880s.

It must be remembered that the 400,000 members in 1890 was a snapshot in time. That number does not include those who previously died, dropped out, re-joined, or joined in the future.

Improved record keeping accurately revealed how many veterans were active members at any time, allowing leaders to create credible promotions to increase membership. In his 1888 address to the National Encampment, Commander-in-Chief John P. Rea proudly reported recent G.A.R. growth with expectations that enrollment would soon exceed 400,000. However, he stressed:

> *Although the growth of our order has been gratifying, and our aggregate membership is large, there are still fully one-half the living honorably discharged soldiers and sailors not upon our rolls.*[3]

Local posts kept a *Descriptive Book* that listed, in chronological order, all veterans who joined the post. Quarterly *Muster Reports* recorded the number of *Members in Good Standing*. Quarterly reports were the basis of official membership numbers. A sampling of *Descriptive Books* and *Muster Reports* showed that two to four times as many veterans joined a post over its life, as were active (good standing) at any point in time. This analysis suggests that aggregate G.A.R. membership may have been more than one million over the life of the organization.

While the late 1880s saw membership climb dramatically, it became difficult to sustain the growth as the 20th century approached. The inevitable march of time took a substantial toll on veterans since by 1900 the average veteran age was about 60 years. It is a testament to dedicated members that the Grand Army did not wither away sooner, as the downward slope in the graph shows a gentle, albeit steady, membership decline.

G.A.R. Structure: Blessing and Curse

The organizational structure of the G.A.R. is both a blessing and a curse for the genealogist and historian seeking information about individual members. Since all records began at the post level and were created by elected officers who personally knew the members by name and face, you can be sure that the information is mostly correct. We are blessed that many recorded entries were written from the veterans spoken words as they told the officer their story—these are valuable original-source records.

The curse stems from the same organizational structure. In 1890, the largest ten departments consisted of an average of more than 400 posts. The largest, Ohio, had 733 posts. The sheer number of posts, many with membership in the low double digits, gave opportunity for records to be lost or destroyed by natural events. Yet we seek these records to help build the story of our ancestors.

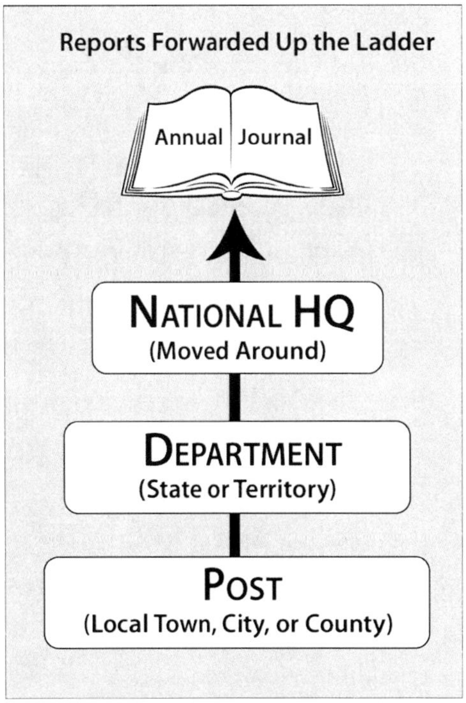

G.A.R. reporting structure with data published yearly in the Annual Journal

There's a Form for That

The G.A.R. functioned similar to any large company or actual army. It was comprised of a National Headquarters, regional offices (department), and local offices (post). Each office reported to the one above. It sounds efficient, but several traits worked against the G.A.R.

1. At every level, the offices were primarily led and staffed by part-time volunteers who may not have had a business background, though national leaders devoted most of their time to the position.
2. The National HQ and many times the Department HQ moved every year to the hometown of the incoming commander and his newly appointed staff, disrupting processes and many times losing records.

There was a form and report for everything, yet the organization structure led to erratic reporting and lost records. In addition, the local post elected new officers each year, resulting in repeating the learning process of reporting yet again. (Initially a District or county organization existed, but this was soon abandoned as it created excessive bureaucracy).[4]

Indeed, the problem of inaccurate reporting was recognized by the G.A.R. The Adjutant General, who was in charge of tracking membership, noted in his report at the 1888 Encampment:

> *I have experienced (perhaps in a diminished degree) the annoyances arising from tardy and irregular reports, upon which all my predecessors have commented; but for an organization so large in numbers and so widely distributed, which habitually changes its entire executive and clerical force once a year, I consider the results achieved exceedingly creditable.*[5]

Records, journals, reports, and other documents are listed in the next pages of this book by the organization level that created or held the records. Detailed reports were made at the post and were forwarded to the department where they were summarized for National Headquarters.

Recognizing Your Veteran

Neither the Union nor the Confederate army used a service or serial number to identify servicemen; only their name was listed. And there were no *dog tags*. The enlistment of millions of men for World War I finally broke the system of tracking service people by their name. In February of 1918, the first service number was issued to a Sergeant Major in the U.S. Army.

With over two million soldiers and sailors serving in the Civil War, duplicate names are frequent. However, when searching nearly every type of service record, including G.A.R. records, a soldier's unit or sailor's ship is usually listed with him. While there are many kinds of 'units' (military organizations), a soldier's *regiment* designation is the most common and dependable identifier to help spot a specific soldier. For sailors, his *ship* is usually listed. Here are a few regiment examples entered on a Muster Report:

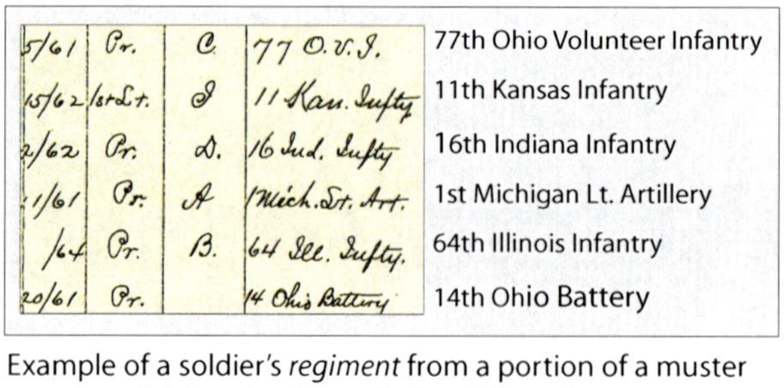

5/61	Pr.	C	77 O.V.I.	77th Ohio Volunteer Infantry
15/62	1st Lt.	I	11 Kan. Infty	11th Kansas Infantry
2/62	Pr.	D.	16 Ind. Infty	16th Indiana Infantry
1/61	Pr.	A	1 Mich. Lt. Art.	1st Michigan Lt. Artillery
/64	Pr.	B.	64 Ill. Infty.	64th Illinois Infantry
20/61	Pr.		14 Ohio Battery	14th Ohio Battery

Example of a soldier's *regiment* from a portion of a muster report, and the author's translation. Prior columns are company, rank, and date entered service.

What Is a Regiment?

The term regiment is used throughout this book, though some readers may not have a clear vision what a regiment is and its role. A regiment is a combat force of approximately 800 soldiers, though that number can vary greatly, usually lower.

In the battlefield, a regiment may be part of a larger organization and battle plan. For example, the largest battles of the Civil War included more than 50,000 Union soldiers. Some of the largest battles and the number of Union troops include:

- Gettysburg 93,000
- Chickamauga 60,000
- Bull Run(2nd) 62,000

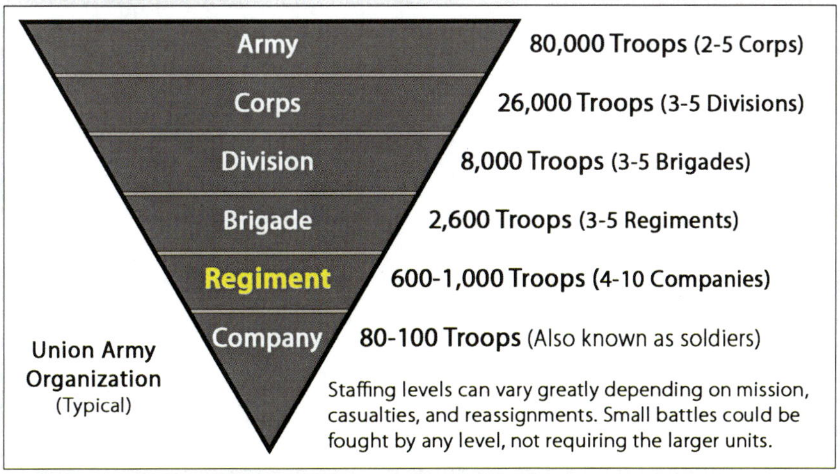

As an example, the Battle of Gettysburg was conducted by a single Union Army, the *Army of the Potomac*, though it was reinforced with seven Corps.

A regiment was a complete fighting force, but not totally independent. They were one of multiple layers of command and organization as the above chart shows.

Regiments were homogeneous organizations; all members stayed together until discharged, killed, or seriously wounded, or until the regiment disbanded. During the Civil War, volunteer regiments were generally recruited from a geographic area encompassing a town or surrounding areas. Most members knew each other as they were neighbors, co-workers, or even relatives. In addition, regimental officers may have been respected leaders of the community.

There were two kinds of service regiments: *Volunteer* and *Regular*. It is important to understand and know this when ordering records from the National Archives.

Volunteer Regiments – These were formed by states as described above and have regimental designations such as *Illinois 56th Infantry*.

Regular Regiments – These units were formed by the U.S. Army. Many of these regiments already existed when the war began, most serving in the west, though additional regiments were formed when the war broke out. A typical *Regular* regiment designation looked like the *10th U.S. Infantry Regiment* (no state reference).

Before You Start Searching

1. Determine the state where your veteran enlisted (or was drafted) and regiment (or ship) he served with.
2. Determine which state your ancestor *lived* in when he might have joined the G.A.R.

All G.A.R. records are organized by the post (state and town) where he joined, NOT where he joined the Army. He may have moved after the Civil War. If you are not sure an ancestor was a Civil War veteran or a G.A.R. member, here are a few easy steps to discover any service.

1. Determine their age during the Civil War. The average age of a Union soldier was in their early 20s. Confederate soldiers were much younger. However, there are instances of boys in their teens enlisting and men in their 50s fighting.
2. Determine where they lived in the early 1860s. Most soldiers served in a state regiment, which were generally organized from an area close to home.
3. Search Ancestry or FamilySearch for your ancestor, in particular for military records. Also, Find a Grave entries may yield military clues, especially a veteran's regiment.
4. Search the *Civil War Soldiers and Sailors System*, a project of the National Park Service: https://www.nps.gov/civilwar/soldiers-and-sailors-database.htm.
5. If you find any records, make a note of the **state, regiment, and company** he served with. This is key to finding other records and eliminating similar names. An entry may look like this: "**Ill. 56th Inf. Reg. Comp. D.**" From the abbreviations, this translates to the "Illinois 56th Infantry Regiment, Company D."

The majority of Union army unit designations follow the above format, although *Infantry* may be replaced with *Cavalry* or *Artillery*. There may be other units, especially Confederate, that appear to be named after someone, who was probably the organizer or leader.

<<

George Gildow's Union Army gravestone shows he served in Ohio's 73rd Infantry Regiment, Company B.

>>

Fountain G. Bristow's Confederate gravestone shows he served in Marmaduke's Arkansas Escort.

> **Research Tip**: Make a note of the *State, Regiment,* and *Company* your veteran served in, or his *Ship* if he was in the Navy. This will help find and identify your veteran in G.A.R. records.

Appendix A includes a list of regiment types, along with example gravestone images.

Where Are the G.A.R. Records?

Before you start searching, gather some general information about a veteran's service, as this will help identify him among millions of records. Note: Determining a veteran's military unit or ship is useful when researching veterans of any era and war.

Harvey County Historical Society, Newton, KS

This small historical society received 15 boxes of G.A.R. and W.R.C. records from the estate of a past W.R.C. post president.

Do not neglect looking in local sources for records.

Physical G.A.R. Record Repositories

Searching physical repositories is the most difficult and time-consuming task, yet this is where the vast majority of G.A.R. records exist. However, the rewards can be worth the effort.

State Archives

Department G.A.R. headquarters were frequently in the state capital and these records eventually became part of state archives. Limited budgets have kept most of these records from being scanned and placed online, though they can usually be researched at the physical repository. Many states have excellent indexes. A state-by-state analysis of these are provided in Chapter Eight.

County and Local Historical Societies & Museums

Some post records are found in a county historical society or museum. Call them to determine the holdings and check their website for catalogs.

City Libraries

Some records may be in a city library, especially if they have a genealogy or special collections department.

Genealogical Societies

Not all genealogy societies have their own libraries or archives, but it is worth a visit or phone call to inquire about G.A.R. holdings.

University Special Collections

Universities in your state or local area may also have received these historical documents and are usually held in their Special Collections.

Research Tip: When talking or corresponding with possible repositories, be clear about what you are looking for. I found that not all staff members are familiar with the term "G.A.R." or confuse it with other organizations. Use the acronym, but also refer to it as the *Grand Army of the Republic, a Civil War veteran organization.*

Digital G.A.R. Records - Online

There are few G.A.R. records online. This is largely due to the scattered nature of the posts and the fact the organization was not part of the U.S. government, therefore records did not find their way to the National Archives. Some states have produced excellent online archives. See Chapter Eight, State Resources, for detailed listings for each state.

Search Tip: Most online services do not include a category for G.A.R. records, they may be scattered among general military categories. In their search boxes, try separate searches for *G.A.R.* and *Grand Army* (No need to spell it out completely).

FamilySearch

FamilySearch is free, though establishing an account is necessary to get the most from it. There are several avenues to finding G.A.R. records, though there is no single database or category that is convenient. All methods start with the *SEARCH* tab.

The *Research Wiki* offers limited records on some states, and these pages are difficult to find. Under *Search*, click on *Research Wiki* then select this and enter *Union Veterans or Lineage Society Records* into its search box. This page will display the most complete listing of G.A.R. records on FamilySearch.

Ancestry

This popular subscription service has a few scattered G.A.R. records. A Card Catalog search reveals only Kansas, New York, and Pennsylvania collections of G.A.R. Post reports and rosters. Kansas and New York are organized by posts, while Pennsylvania records are only listed as Box1, Box2, etc.

Ancestry has many military records for all veterans. A quick search of these may reveal your ancestor's service unit which is useful to know when researching G.A.R. records.

Fold3

Fold3 is owned by Ancestry, yet it is a distinct resource and requires a separate subscription. Many libraries have a Library Edition available to card holders, and the regional National Archive offices provide free access to Fold3.

There are no G.A.R. member or post records in Fold3. A 'Grand Army of the Republic' search will return photographs and images that were sourced from the Library of Congress.

However, there are many Civil War records on Fold3. Detailed access to Compiled Military Service Records provided by the NARA is covered in Chapter Six, Federal Records.

Sons of Union Veterans of the Civil War

The Sons of Union Veterans of the Civil War (SUVCW) still operates from its initial organization. The welcome message on the website states:

> *The Sons of Union Veterans of the Civil War is a fraternal organization dedicated to preserving the history and legacy of heroes who fought and worked to save the Union. Organized in 1881 and chartered by Congress in 1954, we are the legal successor to the Grand Army of the Republic.*

The SUVCW is organized similar to the G.A.R. with a National Headquarters, 29 state *departments*, and local *camps* within those states.

The organization does not possess original or transcribed G.A.R. records. However they provide an excellent U.S. map with links to state pages. A selected state page contains a link to *[Your State] GAR Post and History* and *GAR Records Catalog*. The *GAR Post* webpage lists all known posts within a state and some details of them. The *GAR Records Catalog* lists known record holdings of each post and where they are located.

> **Research Tip**: The SUVCW spreadsheets are the best source of post information and resources:
> **http://suvcw.org/garrecords/**

Additional SUVCW information is provided in Chapter Three, Affiliate Organizations.

Internet Archive

This is a nonprofit that archives many military books and documents. Begin your search with 'G.A.R.', 'Grand Army', or related terms.

A search for Grand Army of the Republic returned over 1,500 items. Specific search terms may return archives you are seeking.

> **Research Tip**: This site archives books, film, websites, audio, and more. SELECT the Book icon from the menu bar before you enter a search term.
> **https://archive.org**

HathiTrust Digital Library

HathiTrust is an excellent source of G.A.R. and Civil War related books, journals, and documents. It is especially useful in finding collections of encampment journals, including state versions.

> **Research Tip**: A search for *Grand Army of the Republic* returned over one million entries. Narrow your search by adding a keyword like *encampment.*
> **https://www.hathitrust.org/**

National Records

While many important records originated at National Headquarters, most are not useful for individual member research unless you are looking for a top officer. Yet, many are fascinating reading as they contribute to the history and understanding of the G.A.R. and its members. Here are the National Records with annotation as to their particular focus, including which publications may be helpful in genealogical research.

Regulations and Operations

Each post was required to purchase a copy from the National Headquarters. Every aspect of managing a post and its activities was covered, with expectations that rules and guidance were to be diligently followed. These included the *Grand Army Blue Book*, *Services*, *Rituals*, and *Manual for the Guidance of the G.A.R.*

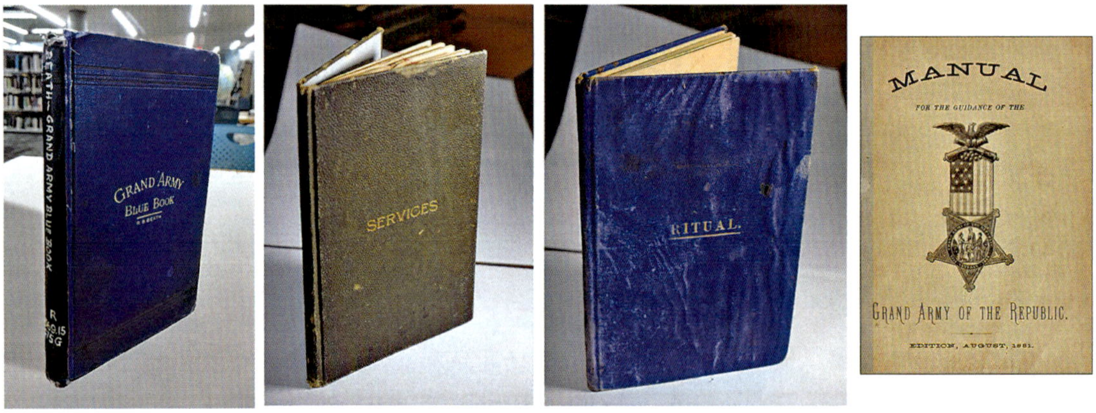

G.A.R. Blue Book, Services, Rituals, and Manual for the Guidance of the G.A.R.

These books covered everything a post needed to conduct business, hold meetings, and produce reports. This is a holdover from military traditions, as the military produces manuals that describes the duties for every waking moment of a soldier's life. These do not contain any names, but help genealogists and historians understand the nature of a veteran's experience with the G.A.R. and the impact it had on them.

The G.A.R. left nothing to chance. The *Ritual* book included sketches of the initiation service that showed the position of each person. Sixteen pages were devoted to the initiation and mustering in of new recruits. This included music, oaths, charge of duties, and many other details, with precise language to be recited.

Exact positions were illustrated for the Sergeant Major (S.M.), Guard (G.), Commander (C.), Quartermaster (Q.M.S), Chaplain (Chap), Officer of the Guard (O.G.), Recruit (R.), and Officer of the Day (O.D.). In addition, their movements and spoken commands are spelled out.

For those not familiar with military terminology, the roles listed come directly from same-named military positions, all in use still today.

Members' roles and positions during initiation service. >

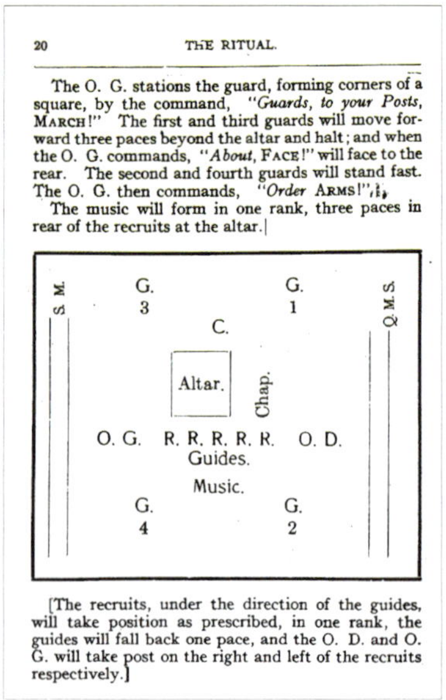
20 THE RITUAL.

The O. G. stations the guard, forming corners of a square, by the command, "*Guards, to your Posts*, MARCH!" The first and third guards will move forward three paces beyond the altar and halt; and when the O. G. commands, "*About*, FACE!" will face to the rear. The second and fourth guards will stand fast. The O. G. then commands, "*Order* ARMS!"

The music will form in one rank, three paces in rear of the recruits at the altar.

[The recruits, under the direction of the guides, will take position as prescribed, in one rank, the guides will fall back one pace, and the O. D. and O. G. will take post on the right and left of the recruits respectively.]

The following publications offer historical looks at G.A.R. practices. Only the annual *Journal of National Encampment* lists individual member information.

National Publications		
Publication	**Research**	**Description**
Journal (Proceedings) of National Encampment	Historical & Genealogy	Annual – Journal of reports and minutes of annual Encampment. Includes many names of Department leaders, attendees, and sometimes recent member deaths.
Grand Army Blue Book	Historical	1855 with Updates – Formally titled *Rules and Regulations*, it soon was called the Blue Book due to its binding.
Manual for the Guidance of the G.A.R.	Historical	1870 w/Updates – Book outlines G.A.R. officer duties.
Rituals of the G.A.R.	Historical	1866 w/Updates – Book with instructions for conducting meetings and initiations.
Services	Historical	Book – Instructions for conducting Memorial Day Services, Burial of the Dead, Memorials, and Installation of Officers.
General Orders	Historical	Letters/Flyers – Contained new regulations or guidance.

Annual National Encampment Journal

Journals of National Encampments are useful for genealogical research. These journals listed national officers, department leaders, committee members, sometimes recent deaths, and usually members who traveled to the event.

81st Congress, 2d Session . . . . House Document No. 387

JOURNAL
OF THE
EIGHTY-THIRD NATIONAL ENCAMPMENT
OF THE
GRAND ARMY OF THE REPUBLIC

INDIANAPOLIS, IND.
AUGUST 28 TO SEPTEMBER 1, 1949

January 3, 1950.—Referred to the Committee on Armed Services and ordered to be printed with illustrations

UNITED STATES
GOVERNMENT PRINTING OFFICE
WASHINGTON : 1949

Journal of the Eighty-Third National Encampment of the Grand Army of the Republic Indianapolis, IND August 29 to September 1, 1949

The journals were titled with different names over the years, but they are easily recognized. Early reports were simply called *Proceedings of National Encampment* with no number, just a year. In 1871 they started numbering journals with titles like *Proceedings of the Fifth Annual Meeting of the National Encampment.*

Early versions were under 50 pages. Later ones grew to over three hundred pages, largely because they included attendee rosters, all past officers, department officers and sometimes members who had died in the previous year. These are fascinating to read and worthwhile to research if you can identify the year(s) your ancestor was in the G.A.R. The last one printed was the eighty-third edition in 1949.

All National Encampment Journals can be found in the *HathiTrust Digital Library,* where numerous collections from various university libraries have been scanned. The home page for this repository is https://www.hathitrust.org. Search for '*Grand Army National Encampment.* and you will find many catalogs. Look for a year or volume you are seeking within a catalog.

What Was an Encampment?

Encampments were formal gatherings of the national organization, a department, or even a post. They were called encampments due to the fact that early versions housed the members in tents, not unlike camps they knew during the war. Later, they eschewed tents for hotels, boarding houses, or even private homes that welcomed veterans during the encampment.

Encampments typically lasted 3-5 days, with business meetings interwoven with festive activities such as parades, and large dinners with dignitaries. Levity, parties, and even rowdy and questionable activities accompanied many encampments.

National Encampments								
Year	No.	Location	Year	No.	Location	Year	No.	Location
1866	1st	Indianapolis, IN	1894	28th	Pittsburgh, PA	1922	56th	Des Moines, IA
1867		No Encampment	1895	29th	Louisville, KY	1923	57th	Milwaukee, WI
1868	2nd	Philadelphia, PA	1896	30th	St. Paul, MN	1924	58th	Boston, MA
1869	3rd	Cincinnati, OH	1897	31st	Buffalo, NY	1925	59th	Grand Rapids, MI
1870	4th	Washington D.C.	1898	32nd	Cincinnati, OH	1926	60th	Des Moines, IA
1871	5th	Boston, MA	1899	33rd	Philadelphia, PA	1927	61st	Grand Rapids, MI
1872	6th	Cleveland, OH	1900	34th	Chicago, Il	1928	62nd	Denver, CO
1873	7th	New Haven, CT	1901	35th	Cleveland, OH	1929	63rd	Portland, MI
1874	8th	Harrisburg, PA	1902	36th	Washington, D.C.	1930	64th	Cincinnati, OH
1875	9th	Chicago, IL	1903	37th	San Francisco, CA	1931	65th	Des Moines, IA
1876	10th	Philadelphia, PA	1904	38th	Boston, MA	1932	66th	Springfield, IL
1877	11th	Providence, RI	1905	39th	Denver, CO	1933	67th	St. Paul, MN
1878	12th	Springfield, MA	1906	40th	Minneapolis, MN	1934	68th	Rochester, NY
1879	13th	Albany, NY	1907	41st	Saratoga Springs, NY	1935	69th	Grand Rapids, MI
1880	14th	Dayton, OH	1908	42nd	Toledo, OH	1936	70th	Washington, D.C.
1881	15th	Indianapolis, IN	1909	43rd	Salt Lake City, UT	1937	71st	Madison, WI
1882	16th	Baltimore, MD	1910	44th	Atlantic City, NJ	1938	72nd	Des Moines, IA
1883	17th	Denver, CO	1911	45th	Rochester, NY	1939	73rd	Pittsburgh, PA
1884	18th	Minneapolis, MN	1912	46th	Los Angeles, CA	1940	74th	Springfield, IL
1885	19th	Portland, ME	1913	47th	Chattanooga, TN	1941	75th	Columbus, OH
1886	20th	San Francisco, CA	1914	48th	Detroit, MI	1942	76th	Indianapolis, IN
1887	21st	St. Louis, MO	1915	49th	Washington, D.C.	1943	77th	Milwaukee, WI
1888	22nd	Columbus, OH	1916	50th	Kansas City, MO	1944	78th	Des Moines, IA
1889	23rd	Milwaukee, WI	1917	51st	Boston, MA	1945	79th	Columbus, OH
1890	24th	Boston, MA	1918	52nd	Portland, OR	1946	80th	Indianapolis, IN
1891	25th	Detroit, MI	1919	53rd	Columbus, OH	1947	81st	Cleveland, OH
1892	26th	Washington, D.C.	1920	54th	Indianapolis, IN	1948	82ND	Grand Rapids, MI
1893	27th	Indianapolis, IN	1921	55th	Indianapolis, IN	1949	83rd	Indianapolis, IN

Railroads offered reduced rates to encampments, so your ancestor may have attended even if the distance seemed excessive. The annual report for 1908 reported that 1,529 members attended the event in Toledo, Ohio.

National encampments moved to a different city each year, as did the department versions, so that as many people as possible could experience the event during their lifetime. Post encampments occurred at a park or venue close to its home.

> **Research Tip**: Look for a National Encampment near your ancestor during the years he may have been a member. The *Encampment Journal* may include his name if he was a delegate or an officer.

These were truly national events. It was a great honor to attend a National Encampment and experience the size, events, and camaraderie of meeting old and new friends. The 1890 Encampment Journal reported about the parade in Boston "*President Harrison and members of his cabinet, with Gen Sherman, occupied a viewing stand on Copley Square.*"

Rock Island train schedule for the 1888 Encampment in Columbus Ohio. (Library of Congress)

44th National Encampment Poster
Atlantic City, N.J.
Sep 19th - 24th, 1910

Attended by 1,622 G.A.R. members, the veterans were greeted by this striking poster. The top portion reveals a 'Sunset at Appomattox' while a ghostly march of soldiers blends into the boardwalk parade. (Library of Congress)

Membership Reports

Membership totals, from annual *National Encampment Journals*, reveal the growth and decline of membership. These statistics were developed from the quarterly reports submitted by posts to departments, where they were consolidated, and then sent to the National Adjutant. The reported numbers were subject to errors and editing, but they are the most accurate available and are useful from the historical research perspective.

A report from the 25th National Encampment Journal in 1891, held in Detroit, Michigan, shows the membership surpassing 400,000 for the first time in the 3rd quarter of 1890.

DEPARTMENTS.	June 30, 1890.		Sept. 30, 1890.		Dec. 31, 1890.	
	Posts.	Members	Posts.	Members	Posts.	Members
Alabama	10	302	10	330	10	332
Arizona	8	300	8	298	8	295
Arkansas	67	1,749	50	1,611	72	2,002
California	117	6.280	100	6,323	107	6,181
Colorado & Wyoming	71	2.663	57	2.719	67	2,940
Connecticut	69	6,805	69	6,897	70	6,946
North Dakota	30	829	16	553	26	731
South Dakota	14	2,095	56	1,882	94	2,783
Delaware	21	1,217	21	1,332	19	1.328
Florida	16	480	16	346	17	414
Georgia	9	407	7	385	9	432
Idaho	17	429	18	447	19	454
Illinois	591	32.355	592	32.568	596	32,641
Indiana	524	25,400	524	25.209	523	25.173
Ind. Ter'y & Oklahoma			20	634	17	605
Iowa	434	20,339	428	20,123	437	20,324
Kansas	484	18.238	478	18,315	480	18.427
Kentucky	130	4.081	109	4.645	136	5.789
Louisiana & Mississippi	17	1,122	17	1.063	17	1,051
Maine	158	9.584	159	9,666	161	9,676
Massachusetts	201	22,460	203	23.458	205	23.592
Maryland	42	2,338	44	2.352	47	2,522
Michigan	373	20,810	362	20,158	378	20,742
Minnesota	165	7.556	145	7.096	176	8,201
Missouri	426	20,317	428	20,732	437	21,028
Montana	18	662	18	675	16	689
Nebraska	252	8,111	251	8,194	255	8,137
New Hampshire	93	5,041	93	5.127	93	5.127
New Jersey	112	7.629	113	7.707	113	7.732
New Mexico	12	340	12	309	10	299
New York	636	39.770	638	40,688	634	40.865
Ohio	712	46,440	725	46,119	733	49,011
Oregon	50	1,698	46	1.774	52	1,905
Pennsylvania	598	44.379	598	44.390	593	43,820
Potomac	13	3,136	14	3.240	14	3.205
Rhode Island	21	2,814	23	2,982	23	2,921
Tennessee	78	3.469	79	3,605	80	3,697
Texas	42	897	38	994	36	1,165
Utah	3	161	3	167	3	188
Vermont	105	5,272	107	5,391	109	5,473
Virginia	40	1,289	41	1,348	42	1,388
Washington & Alaska	60	2,215	58	2,321	59	2,512
West Virginia	82	2,651	80	2,614	82	2,971
Wisconsin	264	13,811	264	13,767	264	13,775
Totals	7,185	397.941	7.138	400.554	7,339	409,489

Membership charts were included in every Encampment Journal beginning in the early 1870s. In sparsely populated areas, more than one state or territory was organized and listed under a single department.

The growth is largely attributable to the push for increased pensions by the G.A.R. which brought many new veterans into the fold. In 1890, the most sweeping pension legislation was passed, providing a pension to all Union Civil War veterans, no disability required.

General Orders

General Orders were official notices informing departments of new regulations or other important communications. Departments were then responsible for distributing the order to posts. One of the most recognized is General Order No. 11, issued on May 5, 1868, by General John A. Logan, Commander-in-Chief declaring May 30 as the day . . .

> *. . . designated for the purpose of strewing with flowers or otherwise decorating the graves of comrades who died in defense of their country during the late rebellion . . .*

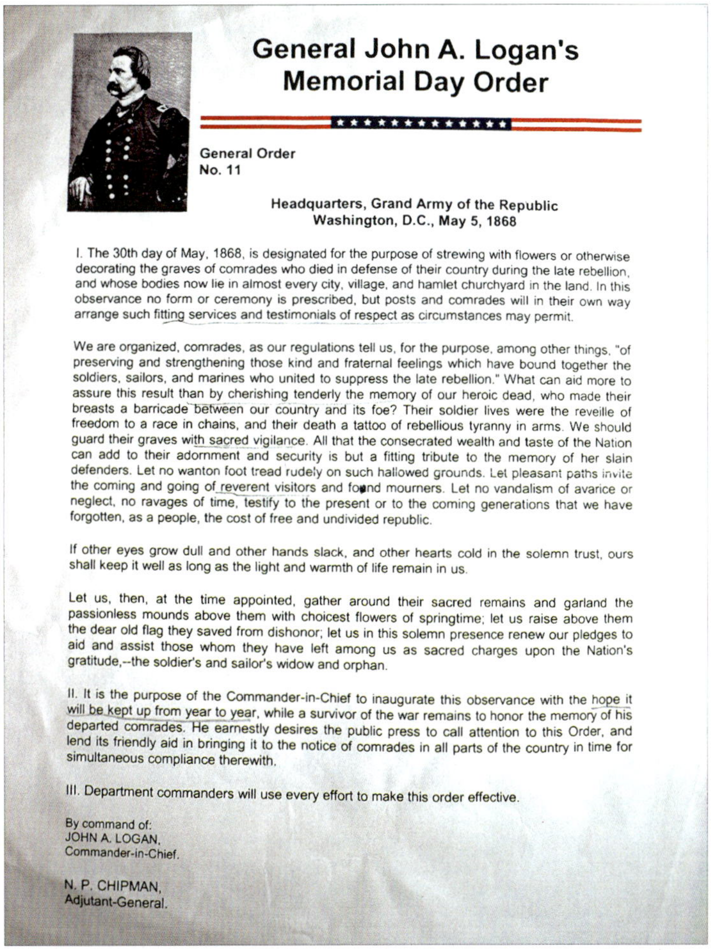

General John A. Logan's Memorial Day Order

General Order
No. 11

Headquarters, Grand Army of the Republic
Washington, D.C., May 5, 1868

I. The 30th day of May, 1868, is designated for the purpose of strewing with flowers or otherwise decorating the graves of comrades who died in defense of their country during the late rebellion, and whose bodies now lie in almost every city, village, and hamlet churchyard in the land. In this observance no form or ceremony is prescribed, but posts and comrades will in their own way arrange such fitting services and testimonials of respect as circumstances may permit.

We are organized, comrades, as our regulations tell us, for the purpose, among other things, "of preserving and strengthening those kind and fraternal feelings which have bound together the soldiers, sailors, and marines who united to suppress the late rebellion." What can aid more to assure this result than by cherishing tenderly the memory of our heroic dead, who made their breasts a barricade between our country and its foe? Their soldier lives were the reveille of freedom to a race in chains, and their death a tattoo of rebellious tyranny in arms. We should guard their graves with sacred vigilance. All that the consecrated wealth and taste of the Nation can add to their adornment and security is but a fitting tribute to the memory of her slain defenders. Let no wanton foot tread rudely on such hallowed grounds. Let pleasant paths invite the coming and going of reverent visitors and fond mourners. Let no vandalism of avarice or neglect, no ravages of time, testify to the present or to the coming generations that we have forgotten, as a people, the cost of free and undivided republic.

If other eyes grow dull and other hands slack, and other hearts cold in the solemn trust, ours shall keep it well as long as the light and warmth of life remain in us.

Let us, then, at the time appointed, gather around their sacred remains and garland the passionless mounds above them with choicest flowers of springtime; let us raise above them the dear old flag they saved from dishonor; let us in this solemn presence renew our pledges to aid and assist those whom they have left among us as sacred charges upon the Nation's gratitude,--the soldier's and sailor's widow and orphan.

II. It is the purpose of the Commander-in-Chief to inaugurate this observance with the hope it will be kept up from year to year, while a survivor of the war remains to honor the memory of his departed comrades. He earnestly desires the public press to call attention to this Order, and lend its friendly aid in bringing it to the notice of comrades in all parts of the country in time for simultaneous compliance therewith.

III. Department commanders will use every effort to make this order effective.

By command of:
JOHN A. LOGAN,
Commander-in-Chief.

N. P. CHIPMAN,
Adjutant-General.

Reprint of *Memorial Day* order distributed to posts.

The actual order did not describe the day as *Decoration Day* or *Memorial Day* (despite the reprint headline above). Members, citizens, and newspapers across the country began using Decoration Day and Memorial Day as they wished. Even the origin of the tradition is of much debate. The G.A.R. was the primary promoter of the practice due to its nationwide influence. Official minutes in the 1872 Encampment Journal referred to this as *Decoration Day*, yet in the 1875 Encampment Journal, *Memorial Day* was used:

> *The services of Memorial Day, by which we have endeavored to commend forever to the tender and grateful recollections of their countrymen, the brave who died, not for themselves, but on behalf of the great cause for which they offered themselves a willing sacrifice, continue to excite the deepest interests.*[6]

Departments Records

Departments were the G.A.R. state headquarters. However, in some areas where populations were small or few posts existed, a department may have covered two states. Departments granted charters to posts in their jurisdiction and acted as the go-between to National Headquarters.

As the country and number of states grew, so did the number of departments. The Homestead Act of 1862 enticed settlers westward and new states filled quickly with veterans after the war. In addition, as pensions became more generous enrollment expanded, and when animosity toward Union veterans in the south subsided, many southern states saw G.A.R. posts founded.

Department Charters

It is important to know when a G.A.R. department was created in your veteran's state. Do not expect to find any records in an area before a charter date. Also, between 1876 and 1881 many departments were *reorganized*, meaning they were probably defunct, and a new charter was issued. Department records prior to a reorganization are scarce.

State Departments					
No.	Department	Original Charter	No.	Department	Original Charter
1	Illinois	April 1, 1866	22	Delaware	January 14, 1881
2	Wisconsin	June 7, 1866	24	Missouri*	May 16, 1867
3	Pennsylvania	January 16, 1867	25	Minnesota*	August 14, 1867
4	Ohio	January 30, 1867	26	Oregon	September 28, 1882
5	New York	April 3, 1867	27	Kentucky	January 27, 1883
6	Connecticut	April 11, 1867	28	W. Virginia*	April 9, 1868
7	Massachusetts	May 7, 1867	29	South Dakota	February 17, 1883
8	New Jersey	December 10, 1867	30	Wash. & Alaska	June 20, 1883
9	Maine	January 10, 1868	31	Arkansas	July 11, 1883
10	California	February 21, 1868	32	New Mexico	July 14, 1883
11	Rhode Island	March 24, 1868	33	Utah	October 8, 1883
12	New Hampshire	June 30, 1868	34	Tennessee	February 26, 1884
13	Vermont	October 23, 1868	35	La. & Miss.	May 15, 1884
14	Potomac	February 13, 1869	36	Florida	June 19, 1884
15	Virg. & N.C.	July 27, 1871	37	Montana	March 10, 1885
16	Maryland*	Jan 8, 1868	38	Texas	March 25, 1885
17	Nebraska	June 11, 1877	39	Idaho	January 11, 1888
18	Michigan*	May 6, 1868	40	Arizona	January 17, 1888
19	Iowa*	September 26, 1866	41	Georgia	January 25, 1889
20	Indiana*	August 20, 1866	42	Alabama	March 12, 1889
21	Colorado & Wyo.	December 11, 1879	43	North Dakota	April 23, 1890
22	Kansas*	December 7, 1866	44	Oklahoma	August 7, 1890

* Between 1876 & 1881 these departments were reorganized.[7]

Department records are difficult to find in many states. Chapter Eight, State Resources, lists known archives of state records and likely places to research. In addition to department-created records, some department collections may also include quarterly post reports, which they summarized and sent to the National Headquarters. Look for the following department records.

> **Research Tip**: Department headquarters and their records were frequently located in state capitals. The state archives or historical society may have these records.

Department Encampment Journal - Annual

Issued after an annual department encampment, the content varies by state. They followed the national format and further listed the active state posts and their officers, encampment attendees, and reports and minutes of the encampment. Many of these are found on HathiTrust Digital Library described earlier, as was this Michigan 1947 journal.

Michigan Journal of 69th Annual Encampment

Post Charter Certificates

The charter-request form, submitted to the department, included the names of all founding members. In return, the actual charter certificate was a very ornate document that included those same names. These charters were supposed to be returned to the department when the post was disbanded.

Miscellaneous Consolidation Forms

The department consolidated reports from posts and sent them to National HQ. These included member counts, financial positions, inventory, and member medical status.

State Rosters - Annual

These may not have been created in all states, and if they were, content varied by state. Typically, these were created in conjunction with annual department encampments.

General Orders - As issued

These are more useful for historical studies of the G.A.R. in your state. They could include re-distributed National Orders and your state's Department General Orders. General Orders might have been endorsements or reminders of upcoming events, or state announcements, including pension information that was important to G.A.R. members.

Post Records

A post charter was granted by the state department; only ten veterans were required to apply for a charter. Counties and larger cities usually included multiple posts, as travel in the late 1800s was not convenient. Even though many small towns in rural areas were only ten miles apart, they still had their own posts.

A January 12, 1885 petition to the Kansas Department requesting a post charter in the town of Wilsey began:

> *We have a number of old soldiers at this place who desire to form a Post of the Grand Army. We are too far from any Post now organized to attend meetings. Our town is 12 miles from Council Grove and about 15 miles from White City.*

Wilsey had but a few hundred residents, as did White City. Council Grove was the largest with a population over 1,000. Still, interest was high even in the smallest towns, but travel by horse or buggy to another town was not convenient and even impossible both for evening meetings during the week and in inclement weather. This was typical of many rural areas across the country, as many states contained hundreds of posts.

Wilsey was granted a charter; their first report filed in March of 1885 reported 20 members.

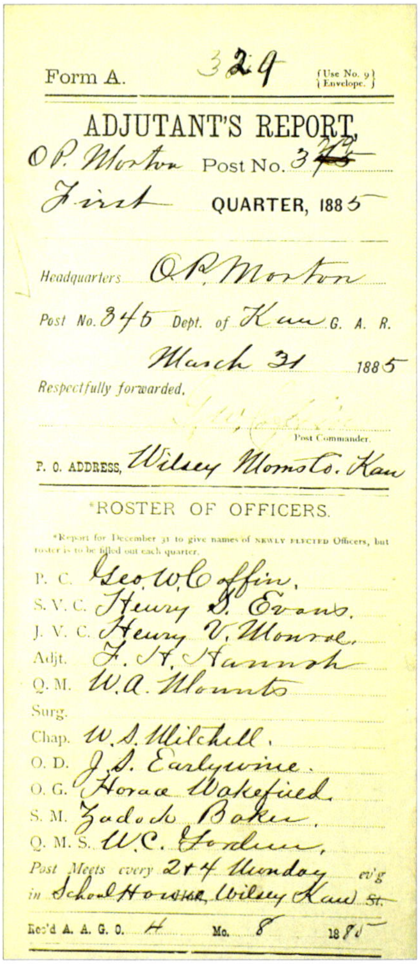
Form A. 329 (Use No. 9 Envelope.)

ADJUTANT'S REPORT,

O.P. Morton Post No. 345

First QUARTER, 1885

Headquarters O.P. Morton

Post No. 345 Dept. of Kan G. A. R.

March 31 1885

Respectfully forwarded,

[illegible]

Post Commander.

P. O. ADDRESS, Wilsey Morris Co. Kan

*ROSTER OF OFFICERS.

*Report for December 31 to give names of NEWLY ELECTED Officers, but roster is to be filled out each quarter.

P. C. Geo. W. Coffin.
S. V. C. Henry D. Evans.
J. V. C. Henry V. Monroe.
Adjt. F. H. Hannah
Q. M. W. A. Mounts
Surg.
Chap. W. S. Mitchell.
O. D. J. S. Earlywine.
O. G. Horace Wakefield.
S. M. Zadock Baker.
Q. M. S. W. C. Gordon.
Post Meets every 2 + 4 Monday ev'g
in School House, Wilsey Kan St.

Rec'd A. A. G. O. 4 Mo. 8 1885

Post Adjutant's Report lists officers, when and where the post met, and other details.

Organized, Chartered, Mustered

While researching posts, you will find the term *organized* and *chartered* used interchangeably. However, the exact meanings are:

- Organized: Date a request for charter was submitted.
- Chartered: Date charter granted by the department.
- Mustered: Date of first post meeting after charter granted.

You may frequently encounter a *re-organized* date. In the 1870s, many posts were disbanded due to loss of members, and their charters lapsed. Subsequent groups of veterans would frequently submit a request for a new charter; hence it was labeled re-organized.

> **Research Tip:** Posts records are the most useful for genealogists and family historians since they contain member details. While the records were created at the local post, much of the information was also transmitted to the department. Look in both places!

Post Reports

Post reports are the most useful records for the genealogists. They contain the greatest amount of member detail, and many were updated quarterly.

Form A – Adjutant's Report and Muster Roll

The Adjutant functioned as the *Secretary*. Form A reveals details of a member's service including age, residence, occupation, unit served, date entered service, discharge date, cause of discharge, and sometimes death information. This was sent to the department every quarter.

Not to be ignored, small details on the Adjutant's Report may lend interesting history to a post story. In this case, the report (previous page) records that the "Post Meets every 2 & 4 Monday" in the "School House, Wilsey Kans."

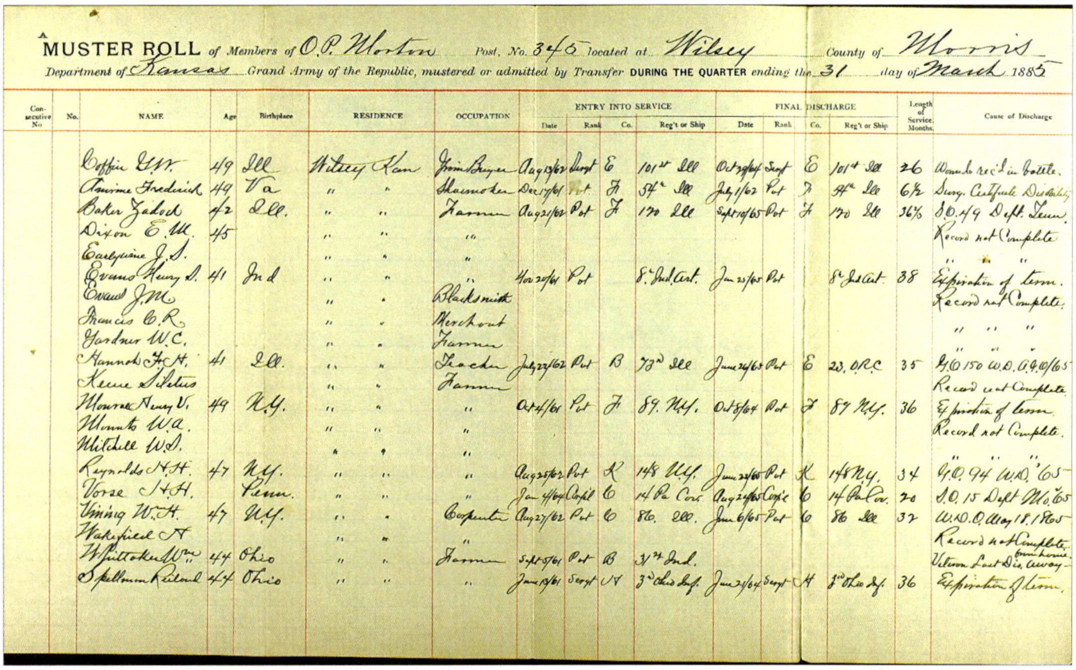

A

MUSTER ROLL *of Members of* O.P. Morton *Post, No.* 345 *located at* Wilsey *County of* Morris

Department of Kansas *Grand Army of the Republic, mustered or admitted by Transfer* DURING THE QUARTER *ending the* 31 *day of* March 1885

Consecutive No	No.	NAME	Age	Birthplace	RESIDENCE	OCCUPATION	ENTRY INTO SERVICE Date	Rank	Co.	Reg't or Ship	FINAL DISCHARGE Date	Rank	Co.	Reg't or Ship	Length of Service, Months	Cause of Discharge
		Coffin G.W.	49	Ill	Wilsey Kan	Grain Buyer	Aug 13/62	Sergt	E	101st Ill	Oct 29/64	Sergt	E	101st Ill	26	Wounds rec'd in battle.
		Cramer Frederick	49	Va	" "	Shoemaker	Dec 17/61	Pvt	F	54th Ill	July 1/62	Pvt	F	54th Ill	6½	Surg. Certificate Disability
		Baker Zadock	42	Ill.	" "	Farmer	Aug 21/62	Pvt	F	130 Ill	Sept 10/65	Pvt	F	130 Ill	36⅔	S.O. 49 Dept. Tenn.
		Dixon E.W.	45		" "	"										Record not Complete
		Earlywine J.S.			" "	"										" " "
		Evans Henry S.	41	Ind	" "	"	Nov 20/61	Pvt		8th Ind. Art.	Jan 25/65	Pvt		8th Ind Art	38	Expiration of term.
		Ewart J.M.			" "	Blacksmith										Record not Complete.
		Francis C.R.			" "	Merchant										" " "
		Gardner W.C.			" "	Farmer										" " "
		Hannah F.H.	41	Ill.	" "	Teacher	July 27/62	Pvt	B	73rd Ill	June 26/65	Pvt	E	23 V.R.C.	35	G.O. 150 W.D. A.G.O./65
		Keene Siletus			" "	Farmer										Record not Complete
		Monroe Henry V.	49	N.Y.	" "	"	Oct 4/61	Pvt	F	89 N.Y.	Oct 8/64	Pvt	F	89 N.Y.	36	Expiration of term
		Monnts W.A.			" "	"										Record not Complete.
		Mitchell W.S.			" "	"										" " "
		Reynolds H.H.	47	N.Y.	" "	"	Aug 25/62	Pvt	K	148 N.Y.	June 22/65	Pvt	K	148 N.Y.	34	G.O. 94 W.D. '65
		Vorse J.H.		Penn.	" "	"	Jan 4/64	Corpl	C	14 Pa Cav.	Aug 24/65	Corpl	C	14 Pa Cav.	20	S.O. 15 Dept Mo. '65
		Vining Wm. H.	47	N.Y.	" "	Carpenter	Aug 27/62	Pvt	C	86 Ill.	June 6/65	Pvt	C	86 Ill	32	W.D.O. May 18, 1865
		Wakefield A			" "	"										Record not Complete.
		Whittaker Wm.	44	Ohio	" "	Farmer	Sept 5/61	Pvt	B	31st Ind.						Veteran Last Dis. away from home.
		Spellman Richard	44	Ohio	" "	"	June 13/61	Sergt	H	3rd Ohio Inf.	June 21/64	Sergt	H	3rd Ohio Inf.	36	Expiration of term.

1885 inaugural Muster Roll for Wilsey, KS O.P. Morton Post #345.

Form B – Quartermaster Report

The Quartermaster functioned as the *Treasurer*. He was responsible for collecting dues, forwarding the dues to the department, and maintaining cash and expenditures. The individual form has little member detail. Note: In 1896 Form A and B were combined into one document, yet still maintained their titles of Form A and B. The combined form also listed the post officers on the front.

Post Rosters

These were not routinely published by most posts but were created for special occasions such as encampments.

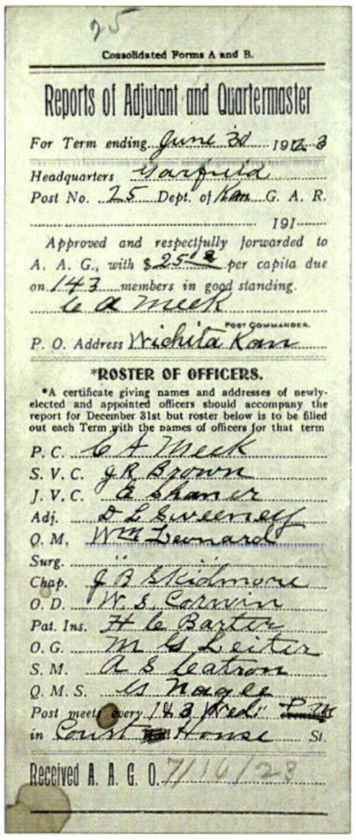

25

Consolidated Forms A and B.

Reports of Adjutant and Quartermaster

For Term ending June 30 1923

Headquarters Garfield

Post No. 25 *Dept. of* Kan *G. A. R.*

191

Approved and respectfully forwarded to A. A. G., with $25¢ *per capita due on* 143 *members in good standing.*

C A Meek

POST COMMANDER.

P. O. Address Wichita Kan

*ROSTER OF OFFICERS.

*A certificate giving names and addresses of newly-elected and appointed officers should accompany the report for December 31st but roster below is to be filled out each Term with the names of officers for that term

P. C. C A Meek

S. V. C. J R Brown

J. V. C. E Shaner

Adj. D L Sweeney

Q. M. Wm Leonard

Surg.

Chap. J B Skidmore

O. D. W. S. Corwin

Pat. Ins. H C Baxter

O. G. M G Leiter

S. M. A S Leatron

Q. M. S. G Nagle

Post meets every 1 & 3 Wed. P.M.

in Court House St.

Received A. A. G. O. 7/16/23

Consolidate Forms A and B,
Adjutant and Quartermaster Report

Descriptive Book

This is one of the more interesting and useful books. The Quartermaster recorded dues payments for each member. This double-wide ledger also included the most extensive information about a member.

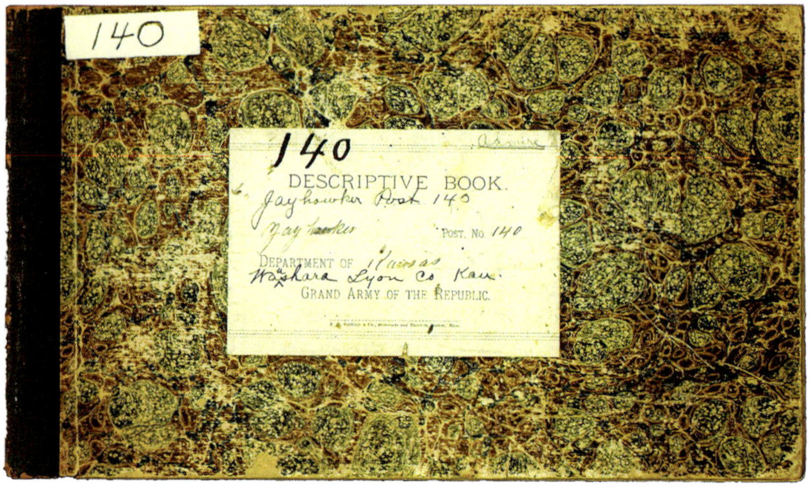

Quartermaster Descriptive Book (Harvey County Historical Society, Newton, KS)

The book provided columns to record a wide range of information, but the practice of collecting and recording the information varied greatly among posts.

Member Data in Descriptive Book	
Name	Length of Service
Age	Cause of Discharge
Birthplace	Date of Muster into G.A.R.
Occupation	When Honorably Discharged
Service Entry Date	When Suspended
Rank	When Dropped
Company, Regiment	When Dismissed
Final Discharge Date	When Reinstated
Rank	Nature of Wounds Received
Company, Regiment	When/What Engagements Wounded

Other Post Records

Post Minute Journals

Minutes are the formal written notes taken during post meetings. The Journal of Minutes can provide interesting reading of the times and can reveal the nature of members and the meetings.

Cash and Receipt Books

This book is a record of income and expenditures.

Post Chaplain's Report

This was not required, but it frequently described post activities such as burials.

Post Medical Descriptive Book

The Post Surgeon was required to provide the department with the information about members' wounds and health status. This information was used in national reports that helped justify pensions. It is hard to find, but when available it can be extremely valuable. It contains description of a veteran's wounds, when they incurred, where, and under what battle conditions. The information can provides great detail to a veteran's story.

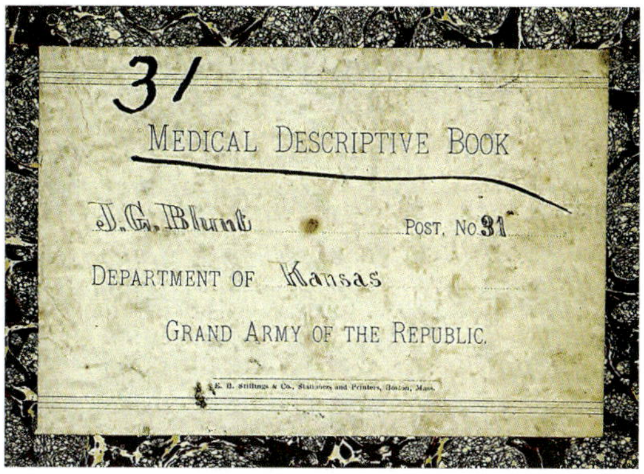

Medical Descriptive Book
Harvey County Historical Society, Newton, KS

Black Book

The Black Book listed veterans who were discharged or barred from the post under dishonorable conditions and not allowed back in.

Post Order Book

This book contained General Orders issued by the ost plus other correspondence.

Personal War Sketches

These are generally rare, yet some posts were diligent in creating them. As the title says, they were stories of the veteran's service experience. They make for fascinating reading if available. Chapter Seven, State Records, covers these in detail.

Encampment Brochures

These are not commonly found or even created for each post or department encampment, but when found they reveal a detailed look at a post during the encampment.

Transfer Cards

This was a certificate issued to a veteran if he wished to transfer to another post, usually due to moving. It is an ornate certificate that gives valuable location data.

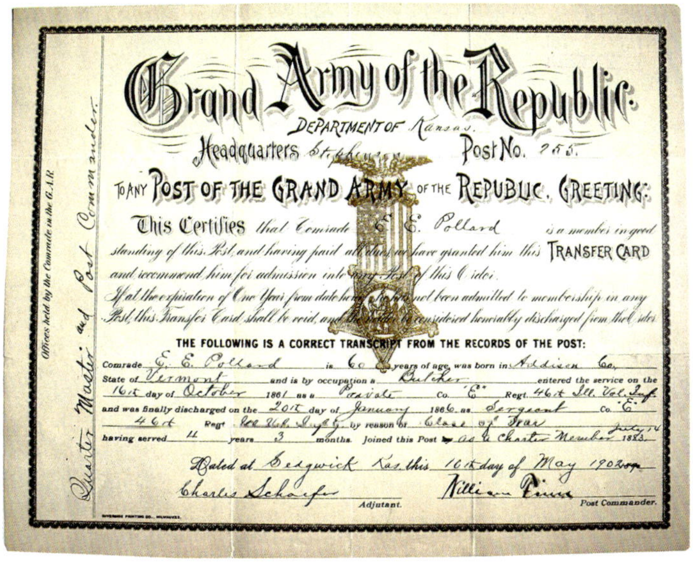

Transfer Card, Harvey County Historical Society, Newton, KS

Most Beneficial Records & Publications (In Order of Usefulness)

A lot of records, journals, reports, and other books have been covered in this chapter. A researcher should not be intimated by the lengthy list of research opportunities. The following list contains the most beneficial and available records generally available for G.A.R. research. If you see any of these, definitely acquire or read them.

- Post Quarterly Muster Roll (Member names, dates, status)
- Post Adjutant's Report Book (New, reinstated, died, and suspended members).
- Post Descriptive and Account Book (Member dues records)
- Personal War Sketches (Personal stories)
- Charter Application (Founding member names)
- Charter Certificate (Founding member names)
- National Encampment Journals of Proceedings (Names and activities)
- Department Encampment Proceedings (Names and local activities)
- Encampment Programs (Organizers and attendees)

Media Resources - Newspapers

Local newspaper reports of Memorial Day activities will frequently list G.A.R. members that participated. Other activities such as veteran funerals, fundraisers, meetings, and encampments were covered by the papers. Finding these accounts can add detail and context to much of your veteran's story.

The National Tribune was the most successful newspaper devoted to veterans and the G.A.R. Published from 1877 to 1911 its banner was "To Care For Him Who Has Borne the Battle, and For His Widow and Orphans." This is a great source for G.A.R. and veteran stories of the time. It is available on the Newspapers.com and GenealogyBank websites.

Newspapers.com

While this resource does not include G.A.R. records and reports, many news stories covered G.A.R. activities. Search newspapers in your area for 'G.A.R.' and separately 'Grand Army'.

https://www.newspapers.com/

GenealogyBank

GenealogyBank is another subscription service that provides access to newspapers.

https://www.genealogybank.com/

G.A.R. Museums

As years progressed in the 1900s, the toll on Civil War soldiers grew, causing posts and departments to disband due to member loss. Records, journals, and reports were sent to government archives, universities, historical societies, and in some cases museums.

Visiting museums dedicated to G.A.R. history can yield new discoveries. There are several impressive museums around the U.S. There also may be smaller ones in your areas or within other organizations. In addition, many of these include displays of the affiliate orders.

Massachusetts

Grand Army of the Republic Memorial Museum
58 Andrew Street, Lynn, MA, 01901
https://garlynn.org/
781-477-7085

Marblehead Museum
170 Washington St.
Marblehead, MA 01945
https://marbleheadmuseum.org/civil-war-museum/
781-631-1768

Michigan

Grand Army of the Republic Memorial Hall & Museum
224 S. Main Street
Eaton Rapids, MI 48827
http://garmuseum.weebly.com/

Pennsylvania

G. A. R. Museum and Library
4278 Griscom Street
Philadelphia, PA 19124
https://garmuslib.org/
215-289-6484

Minnesota

Meeker County Museum at the G.A.R. Hall
308 N Marshall Ave
Litchfield, MN 55355
https://www.meekercomuseum.org
320-693-8911

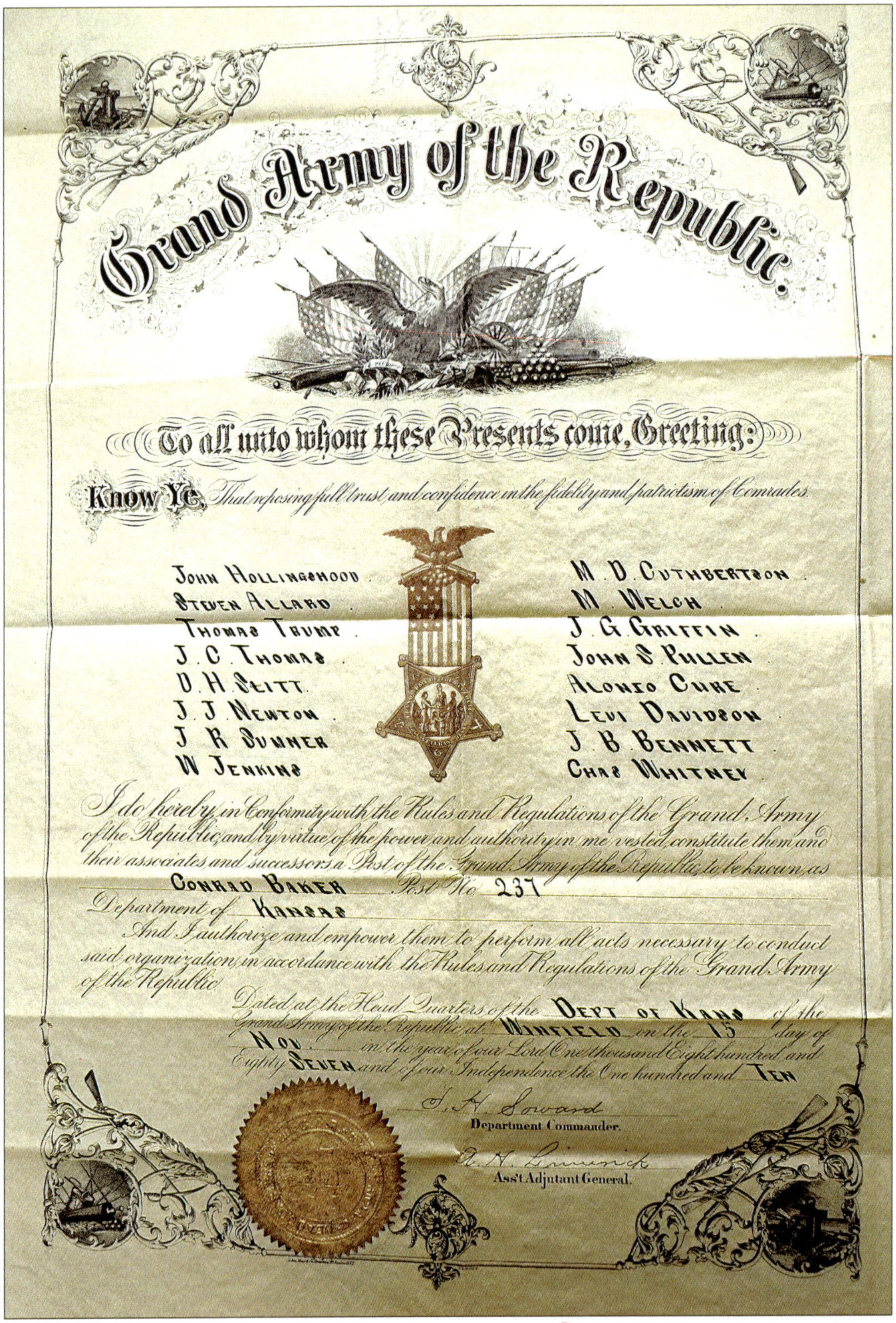

Grand Army of the Republic.

To all unto whom these Presents come, Greeting:

Know Ye, That reposing full trust and confidence in the fidelity and patriotism of Comrades

John Hollingshood.	M. D. Cuthbertson.
Steven Allard.	M. Welch.
Thomas Trump.	J. G. Griffin.
J. C. Thomas.	John S. Pullen.
D. H. Stitt.	Alonzo Cure.
J. J. Newton.	Levi Davidson.
J R Sumner	J. B. Bennett.
W Jenkins	Chas Whitney.

I do hereby, in Conformity with the Rules and Regulations of the Grand Army of the Republic, and by virtue of the power and authority in me vested, constitute them and their associates and successors a Post of the Grand Army of the Republic, to be known as Conrad Baker Post No 237 Department of Kansas

And I authorize and empower them to perform all acts necessary to conduct said organization in accordance with the Rules and Regulations of the Grand Army of the Republic.

Dated at the Head Quarters of the Dept of Kans of the Grand Army of the Republic at Winfield on the 15 day of Nov. in the year of our Lord One thousand Eight hundred and Eighty Seven and of our Independence the One hundred and Ten

S. H. Soward
Department Commander.

A. H. Limerick
Ass't Adjutant General.

Kansas Historical Society

Post #237 Charter, Voltaire, Kansas
November 15, 1887 – Department of Kansas

3

G.A.R. Affiliate Organizations

Womens Relief Corps • Ladies of the G.A.R. • Sons of Union Civil War Veterans • Auxiliary to Sons of Union Civil War Veterans • Daughters of Union Veterans of the Civil War • Affiliate Postcards

Many civilians wanted to help returning Civil War soldiers adjust to a peacetime country. Since the war lasted more than four years and it was fought on America's soil, much of the population had seen the horrors and burdens thrust upon soldiers. Unlike foreign wars, many citizen volunteers tended to the wounded as they returned from battlefields that were sometimes in view of their towns. Yet, the need for assistance did not stop at the signing of peace.

Just as the men of the G.A.R. wanted to honor veterans, preserve their memory, and help those in need—mothers, wives, sons, and daughters of the soldiers, plus grateful citizens wanted to do their part. These efforts were not just praise and parades, but provided direct help overcoming post-war challenges especially to those with injuries. A variety of organizations sprung to life in service to not only veterans, but for widows and orphans as well.

These efforts were not new to post-war America; benevolent organizations began with government sponsored relief organizations during the war.

War-Time America

The type of horrors inflicted on Civil War soldiers had not been seen in decades, if ever in the U.S. Just as the battlefield was not ready and equipped to handle all the casualties, the hospitals, civil authorities, and support organizations back home were not prepared for the onslaught of men needing help. Medical and sanitation facilities were woefully inadequate. More soldiers died from disease, infection, and complications than from immediate battle wounds.

Men were dying on streets of cities and towns; pus-filled wounds and bleeding amputations turned the neighborhoods into repulsive scenes one would expect from the Middle Ages. Substantial help was needed.

The United States Sanitary Commission (USSC) was a civilian-run agency created and funded by donations to support sick and wounded soldiers, supplementing the unprepared and undersized Army Medical Department. The USSC served ten locations located in major metropolitan areas throughout the north.

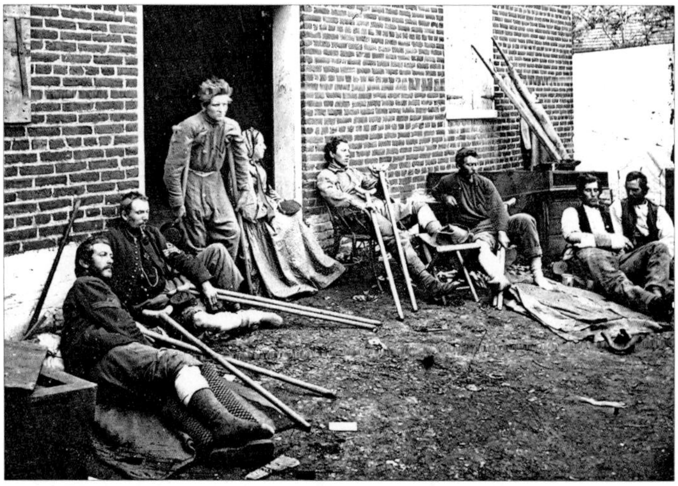

Wounded men at a make-shift aid station outside Fredericksburg, VA
Library of Congress

Another force in helping victims of the war was the United States Christian Commission (USCC). It furnished supplies, medical services, and religious literature to Union troops during the war. Combining religious support with social services, it supplied Protestant chaplains and social workers, and collaborated with the U.S. Sanitary Commission in providing medical services.

USCC facility in Washington D.C. 1865, Library of Congress

Many smaller organizations helped veterans and active-duty soldiers during the war. Most were disbanded or absorbed into larger entities at the end of the war, yet they were the inspiration for groups that became Grand Army of the Republic affiliates.

Transition to Peace

As the 1870s arrived, the country was still in an economic boom period as reconstruction and westward migration fueled expansion. However, soon the country was gripped by the *Panic of 1873*, a banking collapse that quickly led to the *Long Depression*, a financial abyss that would last until the end of the decade—the longest in U.S. history.

Unfortunately, the wounded and sick veterans still existed in vast numbers. In addition:

> *Hundreds of orphans cried for bread; hundreds of women in widows' weeds went about the streets seeking work in the name of soldier husbands, and there was none.*[1]

Immediately after the war, many local and state aid organizations spontaneously cropped up, formed by caring women to provide aid to those in need. Scattered throughout Massachusetts, Ohio, Maine, New Hampshire and most other eastern states, they adopted names such as Soldier's Aid Society, Ladies' Aid Society, Woman's State Relief Corps, and simply Woman's Relief Corps. Many, but not all, were directly associated with local G.A.R. posts, though no national recognition, coordination, or support was in place. By the late 1870s, associations with the G.A.R. at department levels were formed, eventually leading to national endorsements.

Order from Chaos

As with many good ideas, there are frequently competing plans, each with staunch backers. Even the G.A.R. itself went through growing pains due to rival ideas as to the best direction and goals.

Upstart support organizations were not just competitive with each other, some harbored irreconcilable differences, even though they had common objectives. This led to jockeying for preferred positions with the G.A.R.

After years of discussions and sometimes wrangling, by 1883 five official *Orders* were endorsed by the G.A.R.

- Woman's Relief Corps (W.R.C.) 1883
- Ladies of the G.A.R. (LGAR) 1886
- Sons of Union Civil War Veterans (SUVCW) 1881
- Auxiliary to Sons of Union Veterans of the Civil War (ASUVCW) 1887
- Daughters of Union Veterans of the Civil War (DUVCW) 1900

The first acknowledged group was the Woman's Relief Corps, followed by the Ladies of the G.A.R., and then associations of men and women who represented descendants of the veterans. Still, an order of authority was needed. Official designations of *Auxiliary* and *Allied* became important. *Auxiliary* was the most coveted and was bestowed on the W.R.C. while the others were appointed as *Allied* orders.

Woman's Relief Corps (W.R.C.)

After years of fragmented local and state aid societies working towards a common goal of helping the victims of the Civil War, a consolidation was due. During the 15th Annual Encampment in Indianapolis in 1881, a proposal was made to the G.A.R. by the Chaplain-in-Chief to recognize the need for an official women's auxiliary. An appointed committee recommended the following resolutions:

> *Resolved*, That we approve of the project of organizing a National Woman's Relief Corps.
>
> *Resolved*, That such Woman's Relief Corps may use, under such title, the words "Auxiliary to the Grand Army of the Republic."[2]

These resolutions were unanimously adopted; however, under their parliamentary rules, this step did not make the resolutions official. Final adoption did not happen until 1883 while discussion and negotiations continued.

When the Commander-in-Chief issued his call for the 1883 Denver Encampment, he invited the interested auxiliaries, aid societies, and others, to meet at Denver, July 25-26, for the purpose of forming a national association. About fifty women representatives answered the call for their organizations.

W.R.C. Member Badge
Author's Collection

The 1881 report for the organization of a Woman's Relief Corps was presented by the presiding Chaplain-in-Chief, upon whose motion the following resolution was unanimously passed in 1883:

> *Whereas*, We believe it to be our honorable privilege to recognized the magnificent loyalty displayed by the patriotic women of the North during the war of the rebellion, by their loving prayers in our homes, by their contributions to the Christian and Sanitary Commissions, and by their womanly fidelity and devotion on the battle-field and in the hospital; therefore
>
> *Resolved*, That we cordially approve the work which has already been done by the Woman's Relief Corps, and every other organization of loyal women, for the furtherance of the principles of the Grand Army of the Republic.
>
> *Resolved*, That we approve the project entertained of organizing a Woman's National Relief Corps.
>
> *Resolved*, That such Woman's National Relief Corps may use under such title the words "Auxiliary to the Grand Army of the Republic, June 15, 1881."[3]

The Woman's Relief Corps was now the official Auxiliary to the G.A.R. and became crucial to the success of the Grand Army. In some respects, they more effectively achieved the Grand Army's goals. Though positioned as an auxiliary, they were able to raise more money, provide more charity, and promote patriotic education better than the veterans themselves.

Much of the W.R.C. success comes as no surprise due to the nature of membership requirements of each organization. Admittance to the Grand Army of the Republic was limited to Union soldiers and sailors who served from 1861 to 1865. This inherently limited the number of members, dooming the G.A.R. to shrink as age and ailments dispatched older veterans. In contrast, the W.R.C. welcomed into their organization any loyal female over sixteen years of age who showed *good moral character*. A woman did not need to be a wife, mother, daughter, or any other relation to a Union veteran—she simply needed the desire to help.

The G.A.R. reached its peak membership around 1890 with slightly over 407,00 members; The W.R.C. was much smaller, but continued growing for 30 years.

Org	1890	1905	1916	1922
W.R.C.	77,000	126,000	142,000	192,000
G.A.R.	407,000	235,000	136,000	76,000

W.R.C. and G.A.R. membership over the years.
Source: G.A.R. and W.R.C. National Encampment Journals

The W.R.C. overtook G.A.R. membership in 1916 with 142,259 members while the G.A.R. was down to 135,931. This trend continued unchecked due to the toll on aging veterans; the Corps reached its greatest membership in 1922 with 192,363 dedicated women.

The W.R.C. worked vigorously through many programs and causes. Early years concentrated heavily on aid and support for veterans, widows, and orphans while lobbying for pension increases. Their work on Memorial Day ceremonies, cemetery decorations, and monuments are evident today. The W.R.C. was probably responsible for raising much of the $1,000 for this noble statue and cannon mount in the small town of Herrington, Kansas. When the large monument was purchased in 1914, the local G.A.R. post included only 16 members.

In Memory of the "The Boys in Blue" 1861-1866 Erected 1914 By The Herrington G.A.R. and W.R.C. "One Country. One Flag." Photographs by Gary W. Clark

The W.R.C. was also involved in their own efforts to help promote health, charity, and patriotism. Besides building and funding veterans' homes, they created Patriotic Commissions, appointed Patriotic Instructors, and provided flags and Pledge of Allegiance materials to schools. A 1905 resolution instructed members to reach out to their community:

> *The National and Department Patriotic Instructors as far as possible [should] visit summer schools, teachers institutes, and district meetings and distribute patriotic literature and Patriotic Primers.*[4]

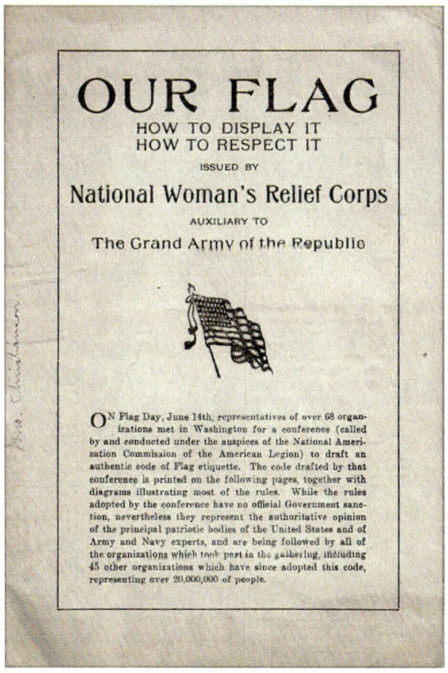

OUR FLAG

HOW TO DISPLAY IT
HOW TO RESPECT IT

ISSUED BY

National Woman's Relief Corps

AUXILIARY TO

The Grand Army of the Republic

ON Flag Day, June 14th, representatives of over 68 organizations met in Washington for a conference (called by and conducted under the auspices of the National Americanization Commission of the American Legion) to draft an authentic code of Flag etiquette. The code drafted by that conference is printed on the following pages, together with diagrams illustrating most of the rules. While the rules adopted by the conference have no official Government sanction, nevertheless they represent the authoritative opinion of the principal patriotic bodies of the United States and of Army and Navy experts, and are being followed by all of the organizations which took part in the gathering, including 45 other organizations which have since adopted this code, representing over 20,000,000 of people.

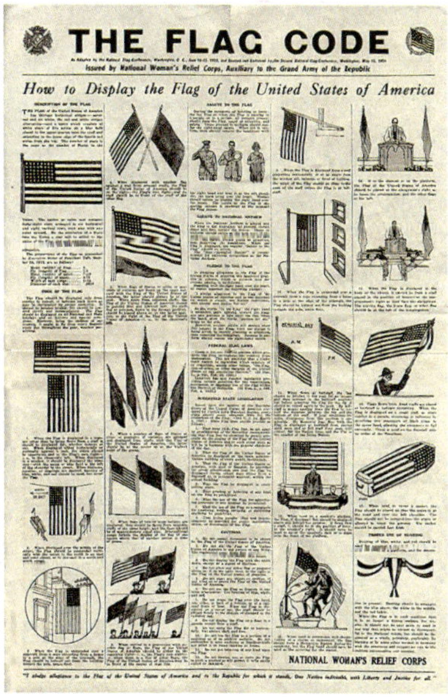

THE FLAG CODE

Issued by National Woman's Relief Corps, Auxiliary to the Grand Army of the Republic

How to Display the Flag of the United States of America

NATIONAL WOMAN'S RELIEF CORPS

A 1924 W.R.C. 12" x 18" pamphlet and Flag Code poster. Author's collection.

While the G.A.R. did not take an official position on suffrage, the W.R.C. did not shy away from the subject and promoted suffrage as the 19th Amendment approached. They approved a 1905 resolution:

> *That the National Woman's Relief Corps indorse (sic) Federal Suffrage for women, believing through this medium much needed work may be performed for veterans, as well as humanity in general.*[4]

This was a bold position as they usually brought little attention to themselves except to honor their own members' graves with markers in cemeteries around the country, many times next to a G.A.R. emblem.[5]

W.R.C. Grave Marker

Eligibility

Of all affiliates, the W.R.C. had the least restrictive membership requirements; no lineage to a veteran was required. A prospective member simply filled out an application with full personal name accompanied by an 'admission fee of not less than one dollar, and the price of badge added. Two current members of the corps needed to recommend the applicant after careful investigation of the good moral character and correct deportment was conducted as outlined in the 1894 *Rules and Regulations for the Government of the Woman's Relief Corps*:

> *Women of good moral character and correct deportment, who have not given aid and comfort to the enemies of the Union, who would perpetuate the principles to which this association stands pledged, and who have attained the age of sixteen years, shall be eligible to membership in the Woman's Relief Corps.*[6]

Member Research

Most Woman's Relief Corps records suffer the same handicap as the lack of centralized G.A.R. archives. Records were created at the *corp* level, corp being the local counterpart to the G.A.R. post, then reports were sent to the state *department*. This record keeping was extensive and comparable to the G.A.R. The challenge is still in finding those records as virtually none have been digitized and made available through internet sources.

> **Research Tip**: Since the W.R.C. was the official auxiliary, a local Corps name and number was the same as their affiliated G.A.R. post.

Corps Records

State archives, local and university libraries, and historical societies are possible holders of original records which include reports and bound ledgers of members, minutes, and accounting. Name-rich ledgers include the *Roster of Woman's Relief Corps* and the *Journal* which captured meeting notes. Unbound reports may include member rosters, registers of national and state convention delegates, and other member reports.

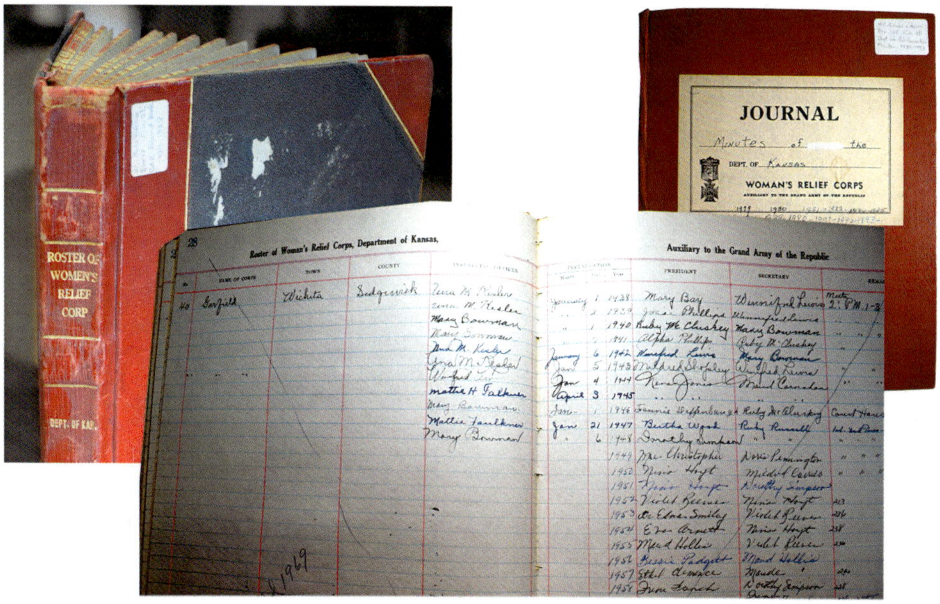

W.R.C. journals can reveal a woman's activities and add new information to their story.
Harvey County Historical Society, Newton, KS

Discoveries sometimes come in bunches. While researching this book, I visited a small town cemetery that held a surprising number of G.A.R. graves and W.R.C. memorials. The well-kept cemetery reflected the town's pride in their history, which included a great historical society located in an equally historic Carnegie Library building. I had visited the society years back to browse their collections, but I was now interested to see if they possessed any G.A.R. or W.R.C. records. This was a long shot, but a calculated idea as there was much local Civil War veteran activity. After explaining my search to the curator, she showed me a catalog of holdings. To my astonishment, their shelves held 15 boxes of records and memorabilia from the local G.A.R. post and W.R.C. corps.

The estate of a patron, who died in 2012, had donated a significant historical collection to the society. That patron's sister, who died years before, was active in the W.R.C. and in possession of local and many state W.R.C. records. While this was a surprising series of events, it can happen anywhere. So look in every possible place for records.

As a bonus, I was reviewing entries and found a page of state W.R.C. Corps Presidents from other cities and towns. A cousin of mine had asked me to keep an eye open for evidence of her paternal grandmother in the W.R.C. Right before my eyes, there was her name as the 1957 President of the Wichita Corps which was in a different county from these records.

A well-worn arch gateway to a section dedicated to Civil War veterans in the Greenwood Cemetery, Newton, Kansas.
Photos by Gary W. Clark

Annual Journals

Just as the G.A.R. held annual conferences at the state (department) and national level, the W.R.C. mirrored those activities. Journals contain transcripts of the proceedings and usually list attendees and officer positions, and many times included previous year member deaths.

Some libraries may have these journals, though these collections are usually spotty. A dependable digital source for these is HahtiTrust Digital Library. This free resource has copies of many national and state conference journals. Here are two collections:

- 1888 - 1923 https://catalog.hathitrust.org/Record/008990712
- 1884 - 1925 https://catalog.hathitrust.org/Record/008990711

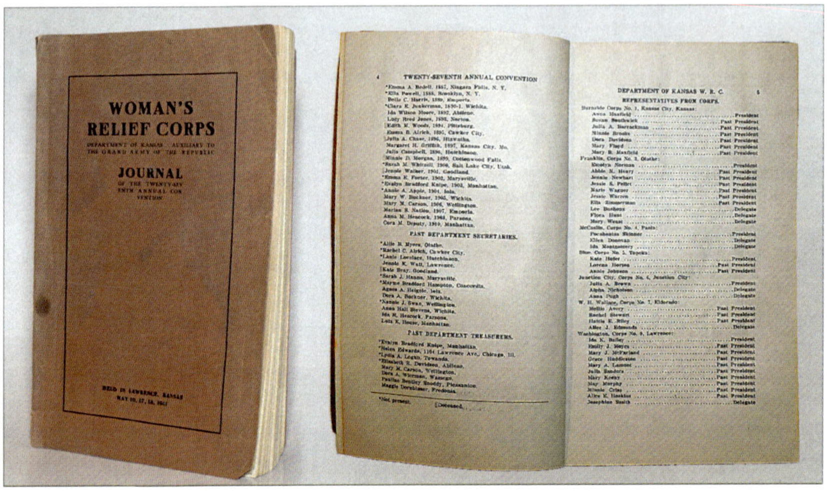

1925 Annual Conference Journal with listings of officers and attendees.

Newspapers

Local newspapers cannot be overlooked when searching for W.R.C. activities and members. No matter the event, oftentime a newspaper report included the names of the women involved, including their officer titles when available. These accounts are great resources to establish where an ancestor was at a specific time and what kinds of activities were included in her life. Using one of the popular newspaper archive services, perform numerous searches using a variety of key words including: 'WRC', 'W.R.C.', and 'Woman's Relief Corps' as the local reporter may have used any of them.

The National Woman's Relief Corps is an active organization. Their website states "A primary goal is to provide perpetually, curation, and preservation of research documents and records that pertain to the GAR as well as the WRC."

Contact

The W.R.C. does not have its own physical facility, though it enjoys space in the Grand Army of the Republic Memorial Museum located in Springfield, Illinois.

Army of the Republic Museum
629 S 7th St
Springfield, IL 62703
https://womansreliefcorps.org/

Ladies of the G.A.R. (LGAR)

The Ladies of the Grand Army of the Republic (LGAR) was a dedicated group of direct descendants of Union veterans. Unlike the W.R.C., the LGAR required its members be wives, mothers, sisters, daughters, or other direct descendants of a veteran. Like the W.R.C., they were formed from a loose coalition of like-minded organizations. But their requirement of a direct relationship to a veteran hobbled their quest to become the lead auxiliary of the G.A.R.

After the 1883 Conference which gave the W.R.C. exclusive title to official Auxiliary, the direct-relation group continued their efforts under the name of Loyal Ladies League. There was occasional bitterness between the two groups, and trouble within the G.A.R. itself as to who should be recognized. In 1886, the Loyal Ladies League met with other organization who already called themselves the Ladies of the G.A.R. and agreed to work together under the name of Ladies of the Grand Army of the Republic (LGAR). Soon, the G.A.R. recognized them as an *Allied* organization, the category created to include additional organizations.

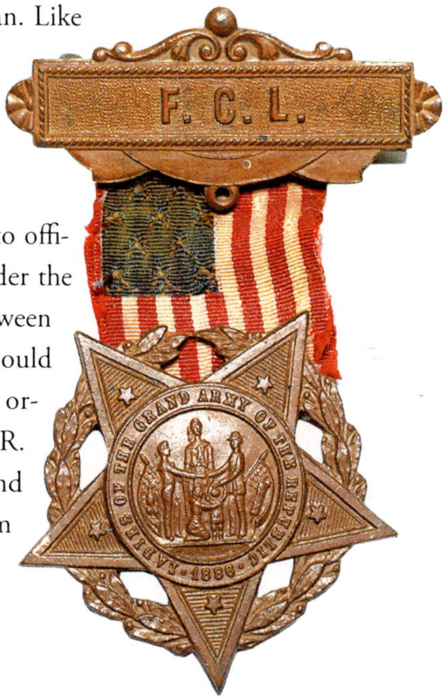

Ladies of the GAR Member Badge
Author's Collection

> **Research Tip:** Local LGAR clubs are called *Circles.* They have a unique name and number different from an associated G.A.R. post.

Eventually the two ladies groups smoothed over their differences and worked alongside each other to the benefit of the G.A.R.

Eligibility

All female blood relatives, ten years of age or over, of honorably discharged Union Soldiers, Sailors and Marines of the Civil War. In addition, ex-army nurses of the war were eligible for membership.

Member Research

Circle Records and Annual Journals

State archives, local and university libraries, and historical societies may hold original records. LGAR records are sparser than W.R.C. records since the LGAR did not reach the size or structural sophistication of the W.R.C. This lack of growth may be attributed to the restriction that members were a blood relative of a Civil War veteran, preventing vast numbers of well-intentioned patriotic women from joining.

Some of their Annual Proceedings are available from Internet Archive and Hathitrust.

- 1921 - 1925 https://catalog.hathitrust.org/Record/008996700
- 1932 - 1942 https://archive.org (search for Ladies of the GAR)

The LGAR is still active, and offers 'Genealogy Look Ups' through its website (hosted by the SUVCW, see below) to help find past members.

Newspapers

Local newspapers may provide the best source of member information if archives cannot be found. Many times, a newspaper report included the names of the women involved.

Contact

The Ladies of the G.A.R. National Headquarters is located in the Ohio Veterans Home.

Ohio Veterans Home, Military Museum
3416 Columbus Avenue
Sandusky, OH 44870. T
http://suvcw.org/LGAR/Home.html (Hosted by SUVCW)

Sons of Union Veterans of the Civil War (SUVCW)

The SUVCW began in 1881 as the Sons of Veterans of the United States of America (SV). Like other G.A.R. affiliates, they originated from the merging of many independent organizations. Initially the SV required members to be a direct descendant of a veteran; however, this would have condemned the organization to a shrinking membership once the aged sons died or became less interested. Eventually, the organization recognized the inevitable and expanded eligibility standards to promote a healthy on-going membership and thus a long-standing society.

SUVCW Member Badge
Author's Collection

In 1925, the SV name was changed to Sons of Union Veterans of the Civil War (SUVCW), under which its federal charter was issued in 1954. The SUVCW is legally recognized as the heir to, and representative of, the G.A.R.

The 1880 *Sons of Veterans Blue Book* documented their constitution, rules, and regulations. Their focus mirrored the *Fraternity, Charity, Loyalty* (FCL) principles of the G.A.R. in paragraphs two through four in Article III:

1. To keep green the memories of our Fathers, and their sacrifices for the maintenance of the Union.

2. To aid the members of the Grand Army of the Republic in the caring for their helpless and disabled Veterans; to extend aid and protection to their widows and orphans; to perpetuate the memory and history of their heroic dead, and the proper observance of Memorial Day.

3. To aid and assist worthy and needy members of our Order.

4. To inculcate patriotism and love of the country, not only among our membership, but among all the people of our land, and to spread and sustain the doctrine of equal rights, universal liberty and justice to all. [7]

Eligibility

The 1880 *Blue Book* defined eligibility for membership as:

> *The sons, not less than eighteen years of age, of deceased or honorably discharged soldiers, sailors or marines, who served in the Union Army or Navy during the Civil War of 1861-65, shall be eligible to membership; also, upon attaining the age of 21 years, all sons of members of the Order in succeeding generations: Provided, That no person shall be eligible who has ever been convicted of any infamous crime, or who has, or whose father has, ever borne arms against the Government of the United States.*[8]

Subsequent changes to membership requirements were less restrictive. Notably, the age requirement is 14 years of age for *Hereditary Membership* and 6 to 14 years for a *Junior Membership*. In addition, descendancy can be through a veteran's brother, sister, half-brother, or half-sister. The SUVCW also created an *Associate Membership* for Civil War enthusiasts that does not require any veteran lineage.

Current Focus

While small in relation to past years, the SUVCW is very active across the country and strives to continue supporting their original goals. It includes 29 Departments (each consisting of one or more states), with over 200 local *Camps*. They participate in Memorial Day programs, community events, and SUVCW activities such as encampments and reenactments.

The SUVCW strives to preserve the history and legacy of Union veterans through several admirable and ongoing projects. There website lists these as:

- National Grave Registration Project: Locate final resting places of all Union Civil War veterans, and establish a database.
- G.A.R. Records Project: Document the location of repositories of records of the Grand Army of the Republic (GAR).
- Monument & Memorial Project: Purpose is the preservation of GAR memorials, including locating and cataloging them.
- Monument & Memorial Grant Fund: The purpose is to assist Camps and Departments in their efforts to refurbish and restore Civil War related monuments and memorials.

Member Research

The SUVCW has undertaken the large task of documenting many aspects of the G.A.R. and Union veterans. It has performed a commendable job of identifying and documenting G.A.R. posts around the country. This author recommends visiting their website when trying to find a G.A.R. post in a particular area.

The SUVCW website (www.suvcw.org) is a helpful source of information. However, finding genealogical relevant information can be confusing (as of early 2021). The place to start is at the PROJECTS & PRESERVATION tab, then click on the *Grand Army of the Republic Records Projects*. Subsequent web pages are of older design and technology, but still valuable.

Research Tip: It is helpful to know the post number you are searching records for. Make note of it from the SUVCW tables described here.

The Records Project web page includes an introduction and a menu bar on the left. Two primary sources of information to explore are the raised buttons on the left labeled **GAR by State** and the **Go Directly to the Catalog.** When selected both the web pages look similar, but contain different useful data.

GAR by State – This displays a U.S. map with all 50 states. Selecting a state will display another page with options – click on [State] **GAR Post & History**. This will display a large table that lists each post, in order, in that state. If you do not know the post number you are looking for, scroll through the entries to identify the town.

This table includes valuable post information including: Location, Post Name, Namesake, Meeting Place, Organized (date), Notes, and Source. Some of this information may give you genealogical clues, or at least back-story information. Note: This page describes the post and its location, and some miscellaneous information under Notes.

GAR Records Catalog – This page shows a similar looking table, but different information. Besides the post name and location, this table describes known post records including: Repository, Location, Collection Name, Description of Items, Medium, Informant, Date, Department, Source. Many locations will reference a state archives, historical society, or other repository. This gives you a place to start searching for records.

Contact

The National Headquarters for the SUVCW is housed at the National Civil War Museum in Harrisburg, Pennsylvania.

1 Lincoln Circle at Reservoir Park
Suite 240 (Nat. Civil War Museum)
Harrisburg, PA 17103-2411
Telephone: (717) 232-7000
http://www.suvcw.org/

A young man in the 1880s wears a SUVCW badge honoring his father, a Civil War veteran. Author's collection.

Auxiliary to Sons of Union Veterans of the Civil War

The National Organization of the ASUVCW is a women's auxiliary to the Sons of Union Veterans of the Civil War. The ASUVCW purpose, as stated on thier website, is:

1. To assist the Sons of Union Veterans of the Civil War in all their principles and objects.
2. To perpetuate the memory of the services and sacrifices of the Union Veterans of the Civil War for the maintenance of the Union, particularly through patriotic and historical observances, especially the proper observance of Memorial Day, Lincoln's Birthday and Appomattox Day.
3. To inculcate true patriotism and love of country, not only among our membership, but to all people of our land, and to spread and sustain the doctrine of equal rights, universal liberty and justice to all. To oppose to the limit of our power and influence, all movements, tendencies and efforts that make for the destruction or impairment of our constitutional Union, and to demand of all citizens undivided loyalty and the highest type of Americanism.
4. To oppose to the limit of our power and influence, all movements, tendencies and efforts that make for the destruction or impairment of our constitutional Union, and to demand of all citizens undivided loyalty and the highest type of Americanism.

ASUVCW Member Badge
Author's Collection

Member Eligibility

Membership in ASUVCW can be obtained through eligibility in several ways as outlined on their website.

1. Women who are lineal or collateral descendants of soldiers, sailors or marines regularly mustered and honorably discharged from the Army, Navy, or Marine Corps of the United States of America during the War of the Rebellion 1861-1865. Daughter, grand-daughter, great-grand-daughter, niece, great-niece, etc., and legally adopted daughters qualify.
2. Mothers, wives, widows, daughters and legally adopted daughters of Sons of Union Veterans of the Civil War in good standing. (Not Associates)
3. Associate Memberships: Available to women who do not qualify through heredity, No. 1 and 2 above, who demonstrate a genuine interest in the Civil War and can subscribe to the purpose and objects of the Auxiliary. An Associate may vote and hold any office except that of National Vice-President and National President. Associates may not exceed one-third of the total membership of the Auxiliary at the time of election.
4. Junior Membership: For young ladies of not less than six (6) years of age.

Contact

The Auxiliary to the SUVCW is a virtual organization with no listed location. They can be contacted at: https://asuvcw.org/

Daughters of Union Veterans of the Civil War

The Daughters of Union Veterans of the Civil War (DUVCW) was an allied organization and is an active lineage society honoring Union veterans. The following information is from the DUVCW web About page:

> *Olive Howard, Harriet Knapp, Eva Merwin, Frank Merwin and Bertha Martin organized the first chapter on May 30, 1885 and held the first meeting on June 3, 1885 at the home of Eva Merwin, 419 First Street, SE Massillon, Ohio.*
>
> *It was incorporated as "The National Alliance of the Daughters of Veterans of the United States of America" on December 12, 1885, in Massillon, Ohio.*
>
> *They were endorsed as an allied order of The Grand Army of the Republic at the 34th G.A.R. Annual Encampment held in Chicago, Illinois, in 1900. The name was changed to Daughters of Union Veterans of the Civil War at the 35th National Convention held in Grand Rapids, Michigan, 1925.*

DUVCW Member Badge
Author's Collection

Member Eligibility

The DUVCW website outlines eligibility in detail:

> *Eligibility for membership in the Daughters of Union Veterans of the Civil War, 1861 – 1865 is based on the eligibility as formerly required for membership in the Grand Army of the Republic and shall be limited to daughters, granddaughters, and great granddaughters of any generation through lineal descent only. As stated in our National Constitutions and Bylaws, this shall never change. The minimum age for membership shall be eight years of age.*

Research

The DUVCW has an informative website with a *G.A.R. Ancestry Gallery* that includes photographs and data for many veterans. Please note, these entries were submitted by members and others, so it is not a comprehensive collection of veterans. However, it is worthy of a visit to see if your ancestor is listed. A soldier entry may include a photo, service unit, and G.A.R. post number if available.

Contact

National Headquarters and Museum
503 South Walnut Street
Springfield, IL
217-544-0616
Email: duvcwnational@comcast.net
Web: https://www.duvcw.org/

Affiliate Postcards

G.A.R. promotion reached new heights in the early 1900s. One of the most popular means of publicizing the G.A.R. and allied orders was the postcard. In 1907, U.S. Postal laws changed to allow the back side of a postcard to contain both the recipient address and a message, freeing the front to display a full artistic image.

Postcards became the social media of the time. The G.A.R., allied orders, and private companies wasted no time creating attractive cards—some revealing a pin-up like appearance.

Woman's Relief Corps teamed with Red Cross.

Ladies of the G.A.R. medal proudly worn.

Sons of Union Veterans of the Civil War
Postcards from author's collection.

Group Portraits

Group portraits became popular in the 1890s with advancements in photography. Many W.R.C. corps, G.A.R. posts, and other organizations seized the opportunity to document their organization.

Woman's Relief Corps, c. 1895 Iowa

Library of Congress

Each corps member proudly wore their membership badge.

4 African-American G.A.R.

Army Organization • Bureau for Colored Troops • Finding Black G.A.R. Members • Valentine Grant Profile

The Grand Army of the Republic was an integrated organization—of sorts. The degree varied greatly across the country, and the extent is not clear in many places. However, there are means to finding Black veterans in the G.A.R. and Army records. First, understanding the condition of the Union Army at the beginning of the Civil War is helpful to your research. Undeniably, when Fort Sumter was fired upon, the Union Army was not equipped, either in man or machine, to fight the looming war.

The Mexican-American War of 1846-1848 established a significant U.S. Army presence along the Mexican border, with more outposts in the upper Plains as settlers and trade routes crossed into traditional Indian lands. Yet, in April 1861, the start of the Civil War at Fort Sumter, the army forces numbered only slightly more than 16,000.

A tremendous effort would be needed to build a fighting force of over 2.2 million Union soldiers. Along the way, attitudes towards Black soldiers changed.

African American Civil War Memorial, Washington, D.C.
Library of Congress

Army Organization

Awareness of the structure of Army commands and units is crucial to finding not only African-American soldiers but any soldiers who served. The G.A.R. records we looked at earlier listed the unit, such as *Illinois 56th Infantry Regiment* (the number may come before the state). The unit helps identify or separates soldiers who had similar names. While there could have been two John Smiths in the 56th, it is still helpful to know the unit, especially when requesting records from archives.

Regiment is the most useful Army identifier after a veteran's name when conducting Civil War research. A common designation in military and G.A.R. records may look like: *Company B, 5th Illinois Infantry.* The 5th Illinois Infantry is a regiment, though many times the word *Regiment* is missing from a recorded entry or a gravestone as it was understood. Reviewing 'What Is a Regiment' on page 21 may be useful. For Navy veterans, *Ship* is comparable to regiment.

Union Army Before the War

At the beginning of the war, the U.S. Army was comprised of only twenty-one regiments; by the end of the war the count was well into the thousands. Individual states provided the majority of regiments. Illinois alone contributed 179 infantry regiments plus more than 60 artillery, cavalry, mounted infantry and independent regiments.

State versus U.S. Regiments

Most soldiers were recruited and trained by their home state; hence, the majority of regiments include the state name. These state regiments eventually came under Federal U.S. Army command, but retained their state affiliation. U.S. regiments did not have state affiliations and were identified with a name like *12th U.S. Infantry;* many are still active today.

As in many wars, the initial sentiment was that the war would not last long. However, as it dragged on, more men and structure were needed to build a winning force. This included a change to the recruitment of and command structure for African-American soldiers.

Bureau for Colored Troops

In May of 1863, the *Bureau for Colored Troops* was organized within the War Department. Some Union states were not interested in forming colored regiments and did not want to integrate white regiments, despite a large number of African Americans that wanted to join the fight. Some were ex-slaves, others were committed to eliminating slavery for their southern brothers—they had a personal stake in the war. The Army needed these volunteers.

Starting a new practice, the U.S. Army recruited them directly into a newly formed regiments labeled United States Colored Troops, commonly abbreviated U.S.C.T. Some state regiments that were already formed around Black soldiers were now converted to U.S.C.T. regiments. For example, the *1st Kansas Colored Volunteers* became the *79th U.S. Colored Troops*.

Come and Join Us Brothers

U.S. Army recruitment poster. Published by the Supervisory Committee for Recruiting Colored Regiments. ca 1863

An exception was made for the *54th* and *55th Massachusetts Infantry* regiments; both were allowed to keep their state designations. The 1989 movie *Glory* was based on the heroics of the 54th in the siege of Fort Wagner, which guarded the Charleston Harbor. They were commanded by a white officer named Captain Robert Shaw, who worked tirelessly for equal pay, equipment, and the same respect white soldiers received. Unfortunately he was killed at the battle for Fort Wagner along with many of his men. Well before the accolades of the movie, many African-American G.A.R. posts carried his name.

Ultimately, about 175 African-American regiments with more than 178,000 free blacks served in the Union Army, bolstering the Union war effort at a critical time. By war's end, men of the U.S.C.T. regiments made up nearly one-tenth of all Union soldiers.

Separate But Unequal

The Bureau for Colored Troops brought efficiency to U.S.C.T. regiments, but not always equitable treatment. Despite objections from Black leaders, the Bureau insisted on assigning white men to commissioned officer positions. Regardless, there were at least eighty-seven African-American officers in U.S.C.T. regiments by the end of the war.[1]

The most glaring problem in the beginning was inequitable pay for Black soldiers. Initially a Black private received $10 a month, with $3 deducted for clothing allowance, while a white private received $13 a month with no deduction. After protests by soldiers and politician of all colors, and the execution of one Black soldier who led a protest, pay was set in June 1864 to $13 a month for all soldiers, of any race, with back pay to January 1.[2]

Identifying Black Regiments

It is helpful to associate a soldier with a specific regiment or ship. This applies to any soldier, of any color. Nearly every archival document of military service and pension activity identifies the veteran with the unit or ship they served with. In the case of researching black units, the regiment is self-identified as an African-American unit by clues in its name.

As we have seen, the designation for black units was U.S.C.T. or United States Colored Troops. That was an overall concept and not the name of a unit, though I have seen examples of that on gravestones. You will find the following common designations for all-Black regiments:

- U.S.C.I. – United States Colored Infantry (Sometimes U.S. Col. Inf.)
- U.S.C.C. – United States Colored Calvary
- U.S.C.A. – United States Colored Artillery (Sometimes H.A. for Heavy Artillery)

Gravestones exhibit a variety of ways unit designations were engraved, but all can be similarly interpreted.

122 U.S.C.I.
U.S. Colored Infantry

17 U.S. Cld. Inf.
U.S. Colored Infantry

3 U.S.C.H.A.
U.S. Colored Heavy Artillery

48th U.S.C.T.
U.S. Colored Troops

Finding Black G.A.R. Members

Integrated G.A.R. posts were not uncommon, though the era still did not extend a welcoming hand to all African Americans. The official stance of the G.A.R. was that a particular religious affiliation, race, or other personal designation was not a requirement for membership, or a reason for exclusion. Even membership applications and subsequent records did not record a veteran's race. Still, a member had to be 'voted in' via a secret ballot by attending members at a meeting. As in other parts of society, discrimination could influence membership, subtly or otherwise.

While identifying black G.A.R. members is not easy, detailed reading of post records can reveal there names. Two army practices help us. First, a soldier's regiment is almost always associated with his name in various records. Second, most black soldiers served in segregated, identifiable regiments. As the Muster Roll below illustrates, you need only to search for an all-Black unit in the Regiment column on G.A.R. forms.

Some paper records, especially G.A.R. records, were written from the spoken word of the veteran. Slang or shortened versions of units may have been uttered. The following 1895 Muster Roll lists two members who served in "U.S. Col," regiments. This muster was obviously written by a single person, most likely the Post Adjutant, and the way he recorded regiments may have been his own style. However, now that we know the different variations and keywords, a researcher can spot an African-American soldier, even in a predominately white post such as this one.

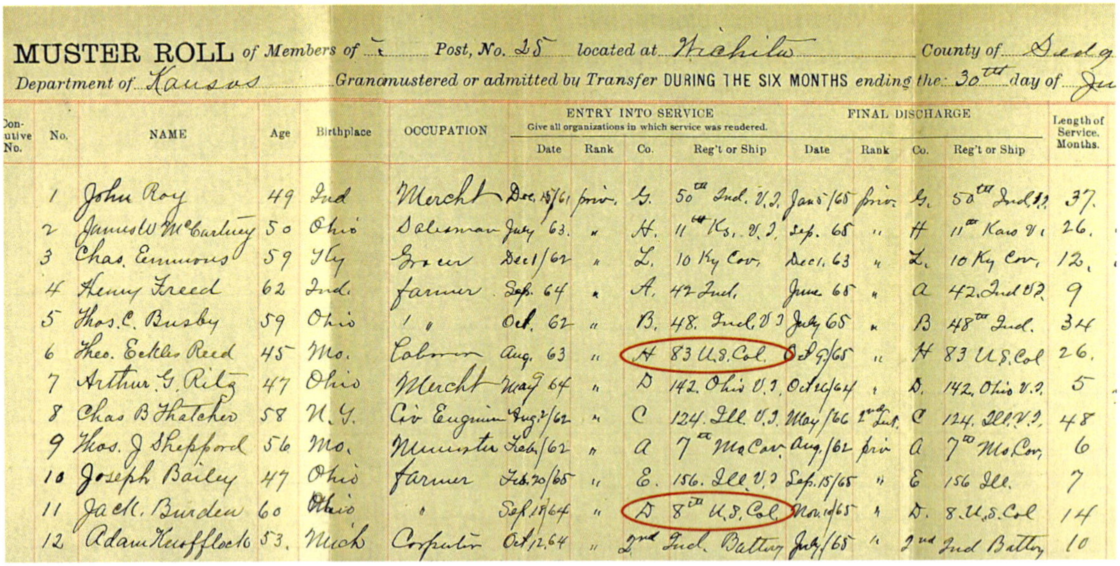

MUSTER ROLL of Members of ... Post, No. 25 located at Wichita County of Sedg...
Department of Kansas Granomustered or admitted by Transfer DURING THE SIX MONTHS ending the 30th day of Ju...

Con-[s]utive No.	No.	NAME	Age	Birthplace	OCCUPATION	ENTRY INTO SERVICE (Give all organizations in which service was rendered.) Date	Rank	Co.	Reg't or Ship	FINAL DISCHARGE Date	Rank	Co.	Reg't or Ship	Length of Service. Months.
	1	John Roy	49	Ind	Mercht	Dec. 15/61	priv.	G.	50th Ind. V.I.	Jan 5/65	priv.	G.	50th Ind. V.I.	37.
	2	James W McCartney	50	Ohio	Salesman	July 63.	"	H.	11th Ks. V.I.	Sep. 65	"	H	11th Kas V.I	26.
	3	Chas. Emmons	59	Ky	Grocer	Dec 1/62	"	L.	10 Ky Cav.	Dec 1. 63	"	L.	10 Ky Cav.	12.
	4	Henry Freed	62	Ind.	farmer	Sep. 64	"	A.	42 Ind.	June 65	"	A	42. Ind V.I.	9
	5	Thos. C. Busby	59	Ohio	"	Oct. 62	"	B.	48. Ind. V.I.	July 65	"	B	48th Ind.	34
	6	Theo. Eckles Reed	45	Mo.	Laborer	Aug. 63	"	H	83 U.S. Col.	Oct 9/65	"	H	83 U.S. Col	26.
	7	Arthur G. Ritz	47	Ohio	Mercht	May 64	"	D	142. Ohio V.I.	Oct 26/64	"	D.	142. Ohio V.I.	5
	8	Chas B Thatcher	58	N.Y.	Civ Engineer	Aug 2/62	"	C	124. Ill. V.I.	May /66	1st Lut.	C	124. Ill. V.I.	48
	9	Thos. J Sheppard	56	Mo.	Minister	Feb /62	"	A	7th Mo Cav.	Aug /62	priv	A	7th Mo Cav.	6
	10	Joseph Bailey	47	Ohio	farmer	Feb. 20/65	"	E.	156. Ill. V.I.	Sep. 15/65	"	E	156 Ill.	7
	11	Jack. Burden	60	Ohio	"	Sep 18/64	"	D	8th U.S. Col.	Nov 10/65	"	D.	8. U.S. Col	14
	12	Adam Knofflock	53.	Mich	Carpenter	Oct 12.64	"	2nd	Ind. Battery	July /65	"	2nd	Ind Battery	10

Muster Roll from the Garfield Post No. 25, in Wichita, Kansas; a predominately white post. However, Comrade Theo. Eckles Reed served in the 83rd U.S. Colored Regiment and Jack Burden in the 8th U.S. Colored Regiment.

There is evidence of new all-white posts being formed once an established post started admitting Black veterans. Twelve years after the above Post No. 25 was organized in Wichita, Kansas, another post was organized. Usually, subsequent posts were formed in an area to address some geographic conveniences; this one met just a few blocks from the first one. Upon inspection of all available records of the second post, no Black veterans were found on their rolls. This could be coincidence, but the evidence is suspicious.

African-American G.A.R. Posts

Researchers may be surprised to learn there were many posts formed by Black veterans. However, it is not surprising that most of these were organized in areas that contributed a substantial number of Black regiments.

While there were many small G.A.R. posts scattered around the Plains and Western states, these areas were typically not populated by many if any Black settlers. So, even though five African American posts were chartered in Kansas—Oklahoma, Nebraska, North and South Dakota, and Texas had none.[3]

Unidentified African American G.A.R. member of Col. John W. Patterson Post #151, Pittsburgh, PA. Library of Congress.

Finding Black Posts

The book *Won Cause: Black and White Comradeship in the Grand Army of the Republic*, by Barbara A. Gannon lists 25 states that included all-Black Posts. Gannon's book, published in 2011, is available from online sources and in many libraries.

The state tables of G.A.R. posts on the Sons of Union Veterans of the Civil War website (http://suvcw.org/garrecords/) also includes a notation on each post that is included in Gannon's book as an African American post.

Number of All-Black G.A.R. Posts - By State					
State	No		State	No
Arkansas	5		Mississippi	16
Delaware	1		Missouri	8
Florida	6		New Jersey	8
Georgia	2		New York	4
Illinois	10		North Carolina	5
Iowa	1		Ohio	12
Indiana	6		Pennsylvania	21
Kansas	5		Rhode Island	1
Kentucky	24		S. Carolina	3
Louisiana	29		Tennessee	9
Maryland	23		Virginia	16
Massachusetts	2		D. C.	3
Michigan	18			

Summarized from *The Won Cause: Black and White Comradeship in the Grand Army of the Republic.*

Valentine Grant Sketch

Valentine Grant served in the 102nd U.S. Colored Infantry, organized originally in February, 1863 as the 1st Michigan Colored Volunteer Infantry Regiment. He enlisted in Marshall, Michigan on January 29, 1864, and fought through South Carolina, Georgia, and Florida before the regiment was disbanded in October 1865.

Grant moved to Kansas in the 1870s and took up farming north of Newton. His gravestone is unique in that the G.A.R. star emblem is engraved into the stone, along with the abbreviated "102nd U.S. COLD. INFT." This was not a government issued gravestone, yet there are three similar gravestones in the Greenwood Cemetery. Not seen in other cemeteries, it was probably carved by a local stonemason.

Valentine Grant, Greenwood Cemetery Newton, Kansas.
Photograph by Gary W. Clark.

5 Badges, Monuments & Cannons

Badges and Ribbons • Buildings as Monuments • Statues and Memorials • Cannons Across the U.S. • Cemeteries as Monuments • Public Relations

The Grand Army of the Republic owed much to its military roots; the detailed reporting, command structure, patriotic dedication, and strict guidance to rules and regulations. Yet one additional military trait was critical to their outward image—a demonstrable sense of *esprit de corps* and the resulting pageantry. Contrary to many secret fraternal societies of the time, the G.A.R. displayed their pride, devotion, and mission in a very public manner.

Whether by design or evolution, the G.A.R. became a marketing machine. They organized or led countless parades, conducted high-profile encampments, proudly displayed their badges, built dramatic structures, and established a permanent presence in cemeteries across the country with grave markers and memorials. Many cemeteries around the country are noted for large Civil War monuments, usually topped with a soldier statue and sometimes flanked by cannons projecting the Union's power.

Following the G.A.R.'s prominence, commercial ventures began associating themselves with the organization. There was G.A.R. Flour, postcards with patriotic images, unofficial souvenir medals, entertainment guides at national encampments which listed the local sporting houses, and a myriad of other trinkets, hats, and adornments. Too bad printed T-shirts were not invented yet, as they would have sold wildly.

Memorial with carved G.A.R. badge.
Soldiers and Sailors Monument, Wichita KS.
Photograph by Gary W. Clark

White bronze 1904 monument flanked by two 32-pounder cannons. The G.A.R. emblem is cast in the monument above "United We Stand." Kingman, KS.

Badges and Ribbons

From the early G.A.R. days, badges identified members at public events and created a sense of pride. While the awarding of badges by the military at the time was not as common as it is now, the practice did gain momentum with the Medal of Honor. On July 14, 1862, President Lincoln signed legislation for the Medal of Honor, the highest U.S. recognition for V*alor Above and Beyond the Call of Duty.*

Soon after the Grand Army was organized, a G.A.R. badge was introduced. First it was basically a shield, but shortly took on a look which lasted through its lifetime. It went through minor iterations, some because it resembled the Medal of Honor too much.

The design was not limited to bronze and cloth badges worn by the members. It became a symbol that graced buildings, monuments, and grave markers.

The eagle symbolizes the country. Its talons grasp crossed cannons and a drawn sword. The cannons represent defense, and the sword is ready to protect the flag.

G.A.R. Member Badge, c. 1880
Author's Collection

Symbolism

The medal was spared no detail; it was designed for maximum imagery. One attempt at symbolism was the medals were to be cast from the brass of captured Confederate cannons. However, this gesture was not scalable, as available Confederate cannons were scarce by 1886.[1] Other symbols were cast in the medal, both front and back. These included a tribute to Fraternity, Charity, and Loyalty (FCL) which are the *Three Guiding Principles* of the G.A.R.

Front

The center was full of meaning. A soldier and a sailor with clasped hands, representing *Fraternity*. Minerva, the goddess of wisdom and warfare, stands tall representing *Loyalty*. A woman and child in the foreground represent *Charity*.

On each side is a flag and an eagle.

The surrounding circle is inscribed **Grand Army of the Republic** on top, the bottom **1861 - Veteran - 1866**.

Outside the center ring, each of the five points include a military arms emblem:

- Crossed Swords: Cavalry
- Crossed Muskets: Marines
- Crossed Cannons: Artillery
- Bugle: Infantry
- Anchor: Navy

Back

Frequent changes were made to the reverse, yet usually minor. This is a c. 1880 version.

Circle of 34 stars represents the states at the beginning of the Civil War.

Inside this ring are the badges or emblems of the Army Corps and Navy.

Outside the center ring, each of the five points varied over time. Early versions had laurels in each place. This one has laurels in the top two positions, then rosettes on the side points, and a campfire in the bottom point.

Officer and Position Badges

Variations of the standard member badge denoted officers of the post, department, or National Headquarters. These included *straps* which replaced the eagle or was placed on the flag between the eagle and the star. Like some military insignia, these variations represented rank or an office, calling attention to the wearer's importance.

If one of these was found among an ancestor's possessions, it was probably a cherished item from his past and an indication of his G.A.R. involvement.

Post Chaplain

Department Chaplain

Post Adjutant, Quartermaster. or Officer of the Day

Officer of the Guard

Senior Officers

Authorized Officer Straps

Commander in Chief	Four silver stars	Sr. Vice Post Commander	Silver leaf
Sr Vice Commander-in-Chief	Three silver stars	Asst. Inspector General	Silver leaf
Jr. Vice Commander-in-Chief	Two silver stars	Aide to Dept Commander	Silver leaf
Department Commander	Two silver stars	Jr. Vice Post Commander	Gilt leaf
Staff of Commander-in-Chief	One silver star	Assistant Inspector	Gilt leaf
Surgeon General	One silver star	Post Surgeon	Gilt leaf
Provision Dept. Commander	One silver star	Council of Administration	Silver "C"
Senior Vice Commander	One silver star	Chaplain-in-Chief	Silver star & cross
Jr. Vice Dept. Commander	One gilt star	Post Chaplain	Two gilt bars
Staff of Dept. Commander	Silver eagle	Post Officer of the Day	One gilt bar
Medical Directors	Silver eagle	Post Quartermaster	One gilt bar
Aides-De-Camp	Silver eagle	Officer of the Guard	Vacant field
Asst Adjutant General to Command-in-Chief	Silver eagle	Quartermaster Sergeant	Chevrons
Post Commander	Silver eagle		

Encampment Badges

Encampments were the G.A.R.'s annual conventions. They were held at the post, department, and national levels. Similar to business conventions, these events wrapped business meetings with social functions and revelry. And like today's conventions, there were promotions, souvenirs, and badges which are highly collectible today, along with the member badges.

Some badges included castings related to the area, photo buttons, or simple silk ribbons. If you find one of these among an ancestor's heirlooms, you can be sure he attended the event. These are samples of the thousands of badges over the years, including national badges, department delegate badges to national encampments, and department encampment badges.

1932 - Springfield, IL
(Dean Palmer collection)

1897 - Buffalo, NY

1896 - St. Paul, MN

1915 - Hutchinson, KS
(Dean Palmer collection)

1914 - Los Angeles, CA

1917 - Salina, KS

Buildings As Monuments

Monuments were not restricted to statues, obelisks, and inscribed plaques. Many buildings, both large and small were the ultimate monument to the G.A.R., and they were functional.

When a post was organized, it needed a place to meet. Early meetings were held in public venues, fraternal halls, and probably a few were held in local establishments where members enjoyed their favorite beverage. As a town's G.A.R. post gained membership, many proudly built their own meeting halls. Large posts in major cities could afford impressive structures. Small rural posts built what was practical and useful.

Library of Congress

The majestic Memorial Hall and G.A.R. building in Rutland, Vermont was built in the 1880s, but fell into disrepair as the veterans dwindled. It was razed in 1931 to make way for a new Federal Building. The town historical society still has the stained glass windows however.

One of the stained glass windows above the entrance includes G.A.R. badge artwork.

Originally a Baptist church, this simple building in Eastport, Maine eventually became the G.A.R. Hall.

A restored Civil War Room still exists in the building today with murals and G.A.R. artifacts as part of the Tides Institute & Museum of Art.

c. 1910 Postcard
Authors collection

G.A.R. Memorial Hall in Hampton, Iowa was built in 1889 to honor 44 Union soldiers from Frankly Countyy who died in the Civil War.

The octagon shaped building includes six stained glass windows paying tribute to the soldiers' roles in the war.
All images this page Library of Congress.

Crossed Bugles

Crossed Rifles

Crossed Swords and Drum

Crossed Cannons

Crossed Swords and Kit

Crossed Sword and Axe

Monuments and Statues

Hundreds, if not thousands of monuments, memorials, statues, and other tributes to Civil War veterans are found coast-to-coast in the largest cities and smallest towns. They are usually in a town square, park, or the area's main cemetery where many were sponsored or championed by the G.A.R. and allied organizations. Fundraising and donations helped finance many pieces while taxes raised funds for others. Some state legislatures authorized county governments to levy property taxes for a specific monument or memorial. This tactic became popular as early as the 1880s and continued into the late 1910s, producing a vast number of iconic structures.

Cuyahoga Co. Soldiers and Sailors Monument
Cleveland, Ohio - Dedicated 1894

Soldiers and Sailors Monument
Easton, Pennsylvania - Dedicated 1900

During the *Gilded Age*, the United States thrived, spending was prolific, and patriotism flourished. During this period (late 1800s to early 1900s), the country was enamored with ancient Egyptian and Greek architecture as a result of increased archeology expeditions in those areas.

G.A.R. growth and influence occurred simultaneously, inspiring it to join the frenzy of building extravagant structures as tributes to veterans and the fallen.

Soldiers and Sailors Monument
New York, NY - Dedicated 1902

Images this page Library of Congress.

Small Town Monuments

Striking monuments were not the exclusive privilege of large cities. Impressive, slightly smaller memorials were erected in towns of only a thousand people. Mass production made this possible, plus much of the stone was quarried locally. Research found this same statue in Wisconsin, Massachusetts, New York, North Carolina, South Dakota, New Jersey, Texas, California, Kansas, and probably many others; sold by the Monumental Bronze Company of Bridgeport, Connecticut.

Cast in White Bronze, a 1904 monument erected in Kingman, Kansas, population 1,785 in 1900.

Drawing from The Monumental Bronze Co. 1882 catalog. Price was $450 for statue only.[2]

ERECTED TO THE
MEMORY OF
THE UNKNOWN DEAD
BY THE W.R.C. No. 71
1911

Peabody, Kansas

A simple acknowledgment of the Eggleston G.A.R. Post describes a memorial in Highland Cemetery, Wichita, KS, dedicated 1897.

All photographs this page by Gary W. Clark.

Cannons Across the U.S.

In addition to countless statues and monuments dedicated to Civil War veterans, cannons were a prized addition to any town square, cemetery, or courthouse grounds. While statues bestowed honor on those who served and died, cannons were a physical expression of the Union's power. Yet, many times, the cannon's origin or its context was immaterial. Just get a cannon.

Many *shore* or *seacoast* guns are found in the Midwest, miles from the nearest sea. But they are big and impressive. Some guns produced after the *War of the Rebellion* frequently share space with Civil War monuments as demand for era-correct cannons exceeded supply. Unfortunately, not all settings were designed by noted architects, as many cannons are exhibited in creative ways; some less attractive or not in a realistic context. Still, they add impact to any display.

Stones River National Cemetery, Murfreesboro, TN (National Parks System).
Resting place of 6,100 Union soldiers, 2,562 are unknown.

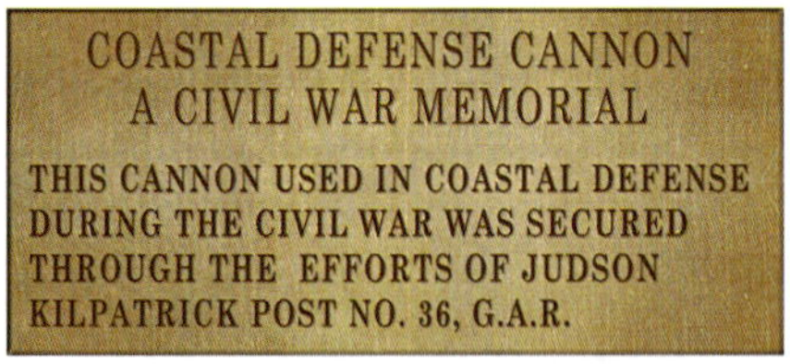

COASTAL DEFENSE CANNON
A CIVIL WAR MEMORIAL
THIS CANNON USED IN COASTAL DEFENSE DURING THE CIVIL WAR WAS SECURED THROUGH THE EFFORTS OF JUDSON KILPATRICK POST NO. 36, G.A.R.

(Partial re-creation of attached placque.)

Dedicated October 4, 1899, the cannon was presented to the town of Newton, Kansas by G.A.R. Post No. 36. Shipped by rail from a Navy Yard in New England, the freight fee was refunded by the railroad as a gesture of respect and patriotism.[3]

Many cannons (and cannon balls) similar to the above were donated to the war effort during World War II and melted for reuse in the military. Cannons such as the above coastal version could weigh 8,000 to 15,000 pounds each, contributing greatly to our country's needs again. As testament to the country, many towns and cities willingly sacrificed their prized veteran monuments to the crisis at the time.

Requested from the Fort Monroe, VA arsenal in 1898, this 8-inch 9,240 pound cannon could project a shell 2.5 miles. It was donated for scrap during World War II. Postcard from author's collection.

In Farragut, Iowa, this 1846 8-inch siege cannon sits in a rebuilt mount from its original stone platform. A unique cannon, reportedly only 18 exist today.[4]

Photo courtesy of Tom Gaard
www.iowacivilwarmonuments.com

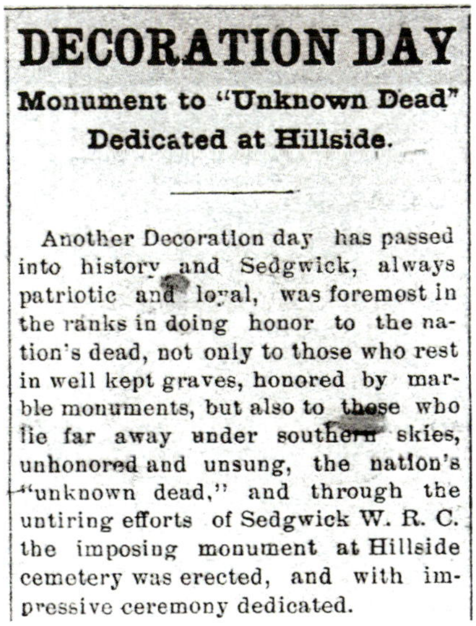

DECORATION DAY

Monument to "Unknown Dead" Dedicated at Hillside.

Another Decoration day has passed into history and Sedgwick, always patriotic and loyal, was foremost in the ranks in doing honor to the nation's dead, not only to those who rest in well kept graves, honored by marble monuments, but also to those who lie far away under southern skies, unhonored and unsung, the nation's "unknown dead," and through the untiring efforts of Sedgwick W. R. C. the imposing monument at Hillside cemetery was erected, and with impressive ceremony dedicated.

The Sedgwick Pantagraph, June 1, 1899

The smallest towns paid respect to Civil War soldiers. In 1900, Sedgwick, Kansas' population was only 622. Still the Woman's Relief Corps raised $132 to transport this cannon from California and build the mount. Photograph by Gary W. Clark.

Cemeteries as Monuments

Few other organizations could turn cemeteries into mass public gathering places, yet that is exactly what happened when the G.A.R. endorsed Memorial Day.

Cemeteries are likely to include statue-topped monuments, pavilions, and other memorials. Still, if an older cemetery does not have such tributes, there is usually a collection of veteran marble gravestones and star-shaped G.A.R. markers.

Fashioned after the membership medal, the bronze markers were created to denote a G.A.R. member, and also functioned as a flag holder for special occasions.

Brass G.A.R. markers are functional as well as attractive. Not all veteran gravestones were the traditional white marble version signifying the person was a Civil War veteran—the G.A.R. star clearly identified a soldier. In addition, the star included a bracket on the back where a flag staff was inserted for special occasions.

The traditional marble veteran gravestone went through numerous changes.

Its size increased over time, particularly its buried depth, creating a more stable monument.

Originally designated for National Cemeteries, laws changed to allow its use in private cemeteries. Many G.A.R. posts conducted surveys of local cemeteries, and ordered stones for soldier graves not marked.

In 1905, the R. H. Graham Post in Moline, Illinois ordered over 110 gravestones from the U.S. Quartermaster-General for Civil War veterans buried in city cemeteries. [5,6]

The inset crest where text is included was only used for veterans serving in the War of 1812, Civil War, and Spanish American War. Stones representing later wars have a smooth face.

G.A.R. Markers

Grave markers went through changes in design; not all were for any documented reason. The familiar five-pointed star marker, fashioned from the G.A.R. member badge, did not appear until the 1880s. This makes sense as the Grand Army became more organized and larger at this time, and somebody came up with a good idea. Slight variations can be found, though most resemble the following styles.

The first version (left) resembled the five-star design in the G.A.R. medal, looking slightly bulky. Soon, a slimmer version (center) was designed which allowed additional detail. Eventually, local production saw post numbers cast into the marker (right). Others with slight variations are found frequently.

Affiliate and Other Markers

Markers other than the five-star G.A.R. version may decorate graves of war veterans and affiliate members. The most common were those placed by the Woman's Relief Corps and the Ladies of the G.A.R. While a separate chapter is devoted to these organizations and others, their grave markers are included here.

Ladies of the G.A.R.

Woman's Relief Corps

Spanish War Veterans

The G.A.R. started a trend that other organizations adopted. As new conflicts occurred, markers representing the Spanish-American War, the Veterans of Foreign Wars, and the American Legion were created to honor veterans. Following this practice, fraternal organizations such as the Masons, Oddfellows, and others created metal markers for their members' graves.

Cemetery, grave, and marker photographs by Gary W. Clark.

Public Relations Machine

Despite some missteps in its early life, by the 1890s the Grand Army of the Republic became an effective public relations machine. They had jettisoned the fiery political rhetoric of the 1870s and concentrated on their three principles: Fraternity, Charity, and Loyalty.

With thousands of posts around the country, each was essentially a public relations office, and each member an agent. Veterans were in every profession, church, and neighborhood. Their presence and stature was something other organizations or businesses could only dream of.

The G.A.R. reinforced their reputation with patriotic activities and heart-tugging reminders of service and sacrifice. Common items like postcards, frequent stories in local newspapers, and their presence in local parades, events, and holidays polished their shine even brighter. Besides enjoying the admiration of adults of all ages, what kid did not like a parade with soldiers bearing arms and flags? An occasional cannon shot was a bonus.

Library of Congress

An 1888 travel poster for the Rock Island R. R. still evoked the adrenalin rush of a battlefield victory.

The c. 1910 masterful postcard of a one-armed veteran and a probable grandchild portrayed a softer emotional message.

These two diverse tactics each reached their respective audiences with undeniable effectiveness.

Author's Collection

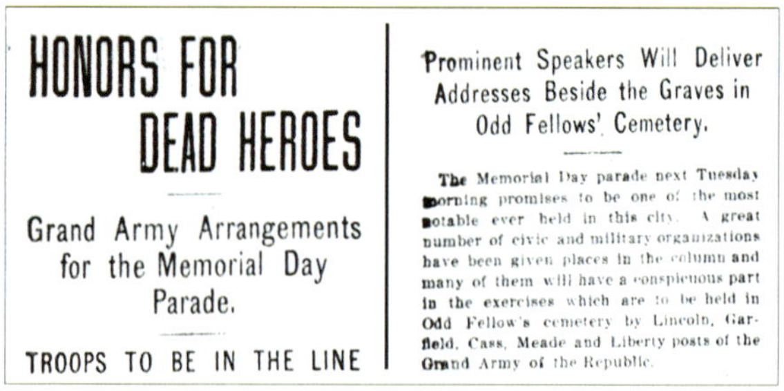

HONORS FOR DEAD HEROES

Grand Army Arrangements for the Memorial Day Parade.

TROOPS TO BE IN THE LINE

Prominent Speakers Will Deliver Addresses Beside the Graves in Odd Fellows' Cemetery.

The Memorial Day parade next Tuesday morning promises to be one of the most notable ever held in this city. A great number of civic and military organizations have been given places in the column and many of them will have a conspicuous part in the exercises which are to be held in Odd Fellow's cemetery by Lincoln, Garfield, Cass, Meade and Liberty posts of the Grand Army of the Republic.

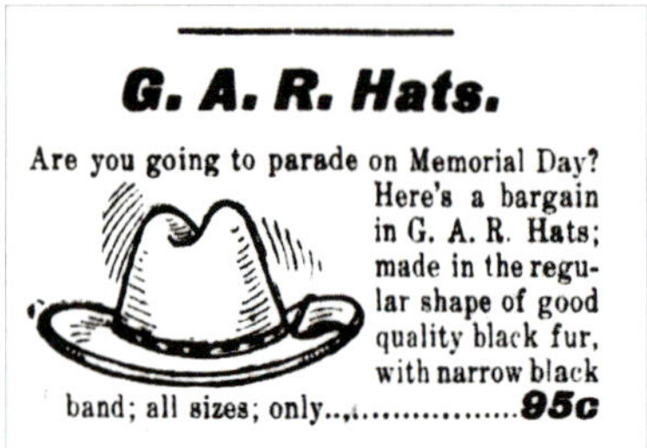

G. A. R. Hats.

Are you going to parade on Memorial Day? Here's a bargain in G. A. R. Hats; made in the regular shape of good quality black fur, with narrow black band; all sizes; only.................**95c**

G.A.R. efforts were sustained by newspapers who covered their efforts and vendors who served the organization.

The *G.A.R. Hats* advertisement appeared in the *San Francisco Examiner* on May 28, 1899, the same day as the above story covering the Memorial Day parade.

Upon Ulysses S. Grant's death in 1885, the G.A.R. immediately released a *Currier and Ives* print extolling the connection of the Grand Army with their *Old Commander.*

Grant was like a *patron saint* to their cause and success. His death presented the opportunity to emphasize the link, especially since this was a time of aggressive pension lobbying.

Library of Congress

National Tribune April 9, 1885

Author's Collection

More than commemorating the last G.A.R. Encampment, this 1948 postage stamp symbolized the coming end of the Grand Army itself with an image of a nearly-drained hour glass, and the juxtaposition of the older veteran with his younger self.

The G.A.R. did not publish its own newspaper or newsletter. However there were several independent veteran newspapers that followed the Grand Army closely.

One popular newspaper was the *National Tribune,* published in Washington D.C. Though it was largely a vehicle for many services and products for veterans, it did report on issues important to all veterans, especially legislation. In fact, the publisher of the paper was the firm of a pension lawyer.

Personal Pride

G.A.R. comrades were proud of their experience and membership in the Grand Army of the Republic; they were easily recognized by their uniforms at national and local encampments. The most popular accessory was the 'slouch' hat with a G.A.R. pin attached to the front of the crown along with a gold braided hat band.

Gathering of Grand Army of the Republic members in 1910 on the U. S. Capitol steps.
Nearly all of them are proudly wearing their slouch hat and G.A.R. badges.
Images Library of Congress

Brass G.A.R. belt buckles are a coveted item among Civil War historians.

David Allen Lambert Collection

6 Federal Records

National Archives • Draft Records • CMSR • Pension Records and Files • National Parks • National Cemeteries • Veteran Headstone Records

While U.S. Government offices did not maintain Grand Army records, they did keep records on many Civil War actions. These may help your G.A.R. search or add details to your veteran's story. Of course, this information is also valuable if your veteran was not a G.A.R. member.

Civil War records and transcribed files reside in various agencies, administrations, and departments; each requiring a different research path. Unfortunately, many of those are not available online. They can only be viewed in person in Washington D.C. or copies made and sent to you.

However, many of the records in the National Archives, though certainly not all, have been digitized and can be found on commercial websites. Companies such as Ancestry, Fold3, FamilySearch, and Archive.org obtained microfilm copies of many record sets and digitized them, making them available through the internet. Still, millions of detailed documents only reside in their original folders from the 19th century in Washington D.C. archive buildings. This chapter describes the most useful records, where they reside, and how to get them.

National Archives and Records Administration

NARA possesses the largest number of historical records. While the administration has responsibility for many regional locations, the Washington D.C. facility, known as Archives I, houses the the vast majority of original Civil War-related records.

Regional NARA Offices are located in:

- Anchorage, AK
- Atlanta, GA
- Boston, MA
- Chicago, IL
- Denver, CO
- Fort Worth, TX
- Kansas City, MO
- Lenexa, KS
- New York, NY
- Philadelphia, PA
- Riverside, CA
- San Francisco, CA
- Seattle, WA
- St. Louis, MO

National Archives Building
700 Pennsylvania Avenue, NW
Washington, DC 20408

Photograph by Gryffindor via Wikimedia Commons
https://commons.wikimedia.org/wiki/File:National_Archives_DC_2007s.jpg

Draft Registration Records

Twenty months into the war, after numerous costly battles including two losses at Bull Run (Manassas) and defeat at the Battle of Fredericksburg in December 1862, the initial belief of a quick Union victory was thoroughly dashed. The unfortunate course of war created the need for more soldiers than were volunteering—a draft was the solution. Congress passed the Conscription Act of March 1863, declaring men between the ages of 20 and 45 subject to a draft, which lasted until the war ended in 1865. Most draft registers survive and provide historical clues about our veterans.

The most available draft information is from the *Consolidated Lists of Civil War Draft Registrations*, that were created from state registers by the Provost Marshal General Bureau. The actual draft registration registers are available in NARA regional archives and sometimes contain more information than the consolidated lists. The draft registers were divided into three parts, or classes; and were collected by state, and congressional district.

- Class I - Men between the ages of 20 and 35 subject to military duty and unmarried men above 35 and under 45 subject to military duty
- Class II - Married men above 35 and under 45
- Class III - Volunteers

Information recorded on the consolidated lists include:

- Residence
- Name
- Age 1st July 1863
- White or Colored
- Profession/Occupation
- Married/Unmarried
- Place of Birth
- Former Military Service

The following statement from NARA describes access to the records:

> *This record has been digitized and made available on-line by our partner, Ancestry.com, for a fee. The digitized records on Ancestry.com are available free of charge in all NARA Research Rooms, including those in our regional archives and Presidential Libraries. (https://catalog.archives.gov/id/4213514)*

Also, many public libraries offer Ancestry *Library Edition* free of charge.

Record Access

Ancestry.com **https://www.ancestry.com/search/collections/1666/**

Research Tip: The National Archives may be called numerous names, including:

- National Archives and Records Administration
- NARA
- National Archives
- Archives.gov

Compiled Military Service Records (CMSR)

The Compiled Military Service Record (CMSR) is one of the most extensive collections of Union veteran information available. It may tell you the battles he was in, where he was at different times, his pay, his wounds if any, and other items.

The CMSR is not a single record, but a jacket or folder of records that were transcribed from detailed original documents. Each soldier has a Compiled Military Service Record for each regiment in which he served and contains basic information about the soldier's military career. From NARA:

> *Beginning in the 1890s, the War Department created the Compiled Military Service Record (CMSR) to document the military service of soldiers. Transcribed from original muster and pay rolls, regimental returns, descriptive books, hospital rolls, and other records, the CMSRs were intended to permit more rapid and efficient checking of military and medical records in connection with claims for pensions and other veterans' benefits.*[1]

A CMSR file may include dozens of documents. Images obtained of one of the author's ancestor's CMSR held 27 records including *Volunteer Enlistment* papers, *Muster Rolls*, *Memorandum From Prisoner of War Records*, *Muster-out Roll* including pay and bounty records, *Descriptive Roll*, and others.

Record Access

Though some CMSR records are available on Fold3, most need to retrieved from the National Archives. Some basic information for a veteran is needed to request records from NARA.

Research Tip: Gather veteran information before you start requesting records. The request forms have **Required Minimum** and **Additional Information** fields. Use research methods in Chapter 2 to unearth as much information as possible.

Required	Additional
Name	Unit Served (Regiment)
State Served	Arm of Service**
War Served In	Officer or Enlisted Man
Union or Confederate	Date of Birth
Volunteer or Regular*	Place of Birth
	Date of Death
	Place of Death

* Volunteer or Regular Regiment. See Ch. 2 section "What is a Regiment?"
** Arm of Service: Infantry, Cavalry, Artillery (May include modifiers such a "*Light* Artillery," "*Mounted* Infantry," etc.)

Retrieving CMSR files requires more work than simply downloading records from the untold millions. However, there are three methods by which you can acquire them.

1. Order copies from the National Archives.
2. Hire a service that specializes in gaining/making copies.
3. Visit the National Archives and make copies yourself.

None of these are quick, and some incur moderate fees. Note: Access to the records may be restricted by closure of the National Archives. This can be determined easily by visiting its website: **https://www.archives.gov**.

Order Copies – Begin with the NARA web page titled *Requesting Copies of Older Military Service Records*, which covers pre-World War I records:

https://www.archives.gov/veterans/military-service-records/pre-ww-1-records

You are given two choices for ordering, one via an on-line request, the other is submitting a paper form (NATF 86) through the mail. Note: If you choose to order on-line, you will be required to create an account with National Archives, which is simple and free. There is a copy fee, starting at $80.00 for a file up to 100 pages. There is no charge if records cannot be found.

Hire a Service – A convenient way to search for records and receive copies is to hire a service that specializes in National Archives records. They will go to the National Archives, get the records and make digital copies themselves. You will still need to provide them as much information as you know; using the information requirements when ordering from the National Archives is a good guideline.

Several services may be available. This author had success with a company named *American Civil War Ancestor* (**www.acwancestor.org**). They visit the National Archives and *pull* the records and make digital copies. Their years of work (I used them in 2013) gives them expert knowledge in finding records. I recommended them to a fellow genealogist in 2019 and their performance was again outstanding.

Visit the National Archives – You should be thoroughly prepared before you start a research journey, including gathering the veteran information outlined above. Also, there are NARA guidelines for visitation restrictions, registering for research, and requesting records. Become thoroughly familiar with their guidelines listed on *Why Visit the National Archives* and *Plan Your Visit web* pages:

https://www.archives.gov/locations/why-visit
https://www.archives.gov/research/start/plan-your-visit

> **Research Tip:** Before attempting to retrieve specific CMSRs in-person, check if they are available on Fold3. The NARA staff will not pull records that are already digitized and accessible online. See next section titled "Online CMSR Records"

A good digital camera or smart phone is the best method to make copies. Practice copying documents before you go, and capture high-resolution images (settings in camera or phone). You will not be allowed to use a flash so learn how to turn it off.

Typical records in a CMSR file.

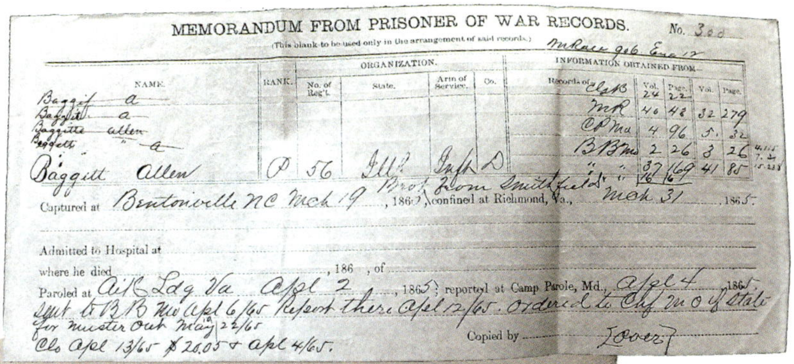

MEMORANDUM FROM PRISONER OF WAR RECORDS. No. 300

(This blank to be used only in the arrangement of said records.)

NAME.	RANK.	No. of Reg't.	State.	Arm of Service.	Co.	Records of	Vol.	Page.	Vol.	Page.
Baggitt Allen	P	56	Ill	Inft	D					

Captured at Bentonville NC Mch 19, 1865, confined at Richmond, Va., Mch 31, 1865.

Admitted to Hospital at

where he died, 186, of

Paroled at Aik Ldg Va Apl 2, 1865; reported at Camp Parole, Md., Apl 4, 1865.

Copied by [over]

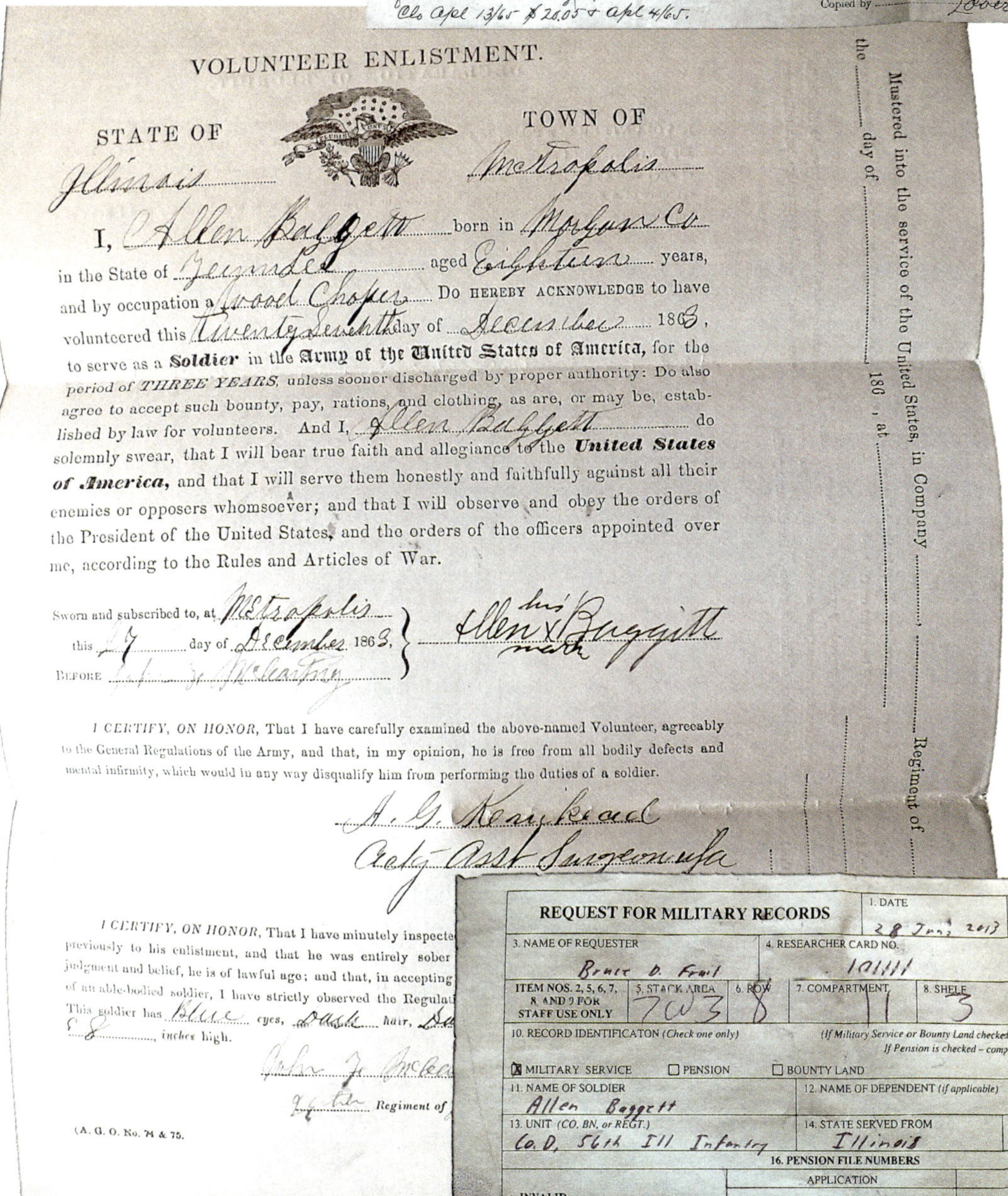

VOLUNTEER ENLISTMENT.

STATE OF Illinois TOWN OF Metropolis

I, Allen Baggett born in Morgan Co in the State of Tennessee aged Eighteen years, and by occupation a Wood Chopper Do hereby acknowledge to have volunteered this Twenty Seventh day of December 1863, to serve as a Soldier in the Army of the United States of America, for the period of THREE YEARS, unless sooner discharged by proper authority: Do also agree to accept such bounty, pay, rations, and clothing, as are, or may be, established by law for volunteers. And I, Allen Baggett do solemnly swear, that I will bear true faith and allegiance to the United States of America, and that I will serve them honestly and faithfully against all their enemies or opposers whomsoever; and that I will observe and obey the orders of the President of the United States, and the orders of the officers appointed over me, according to the Rules and Articles of War.

Sworn and subscribed to, at Metropolis this 27 day of December 1863, Before — Allen his x mark Baggitt

I CERTIFY, ON HONOR, That I have carefully examined the above-named Volunteer, agreeably to the General Regulations of the Army, and that, in my opinion, he is free from all bodily defects and mental infirmity, which would in any way disqualify him from performing the duties of a soldier.

A. G. Kincaid
Actg Asst Surgeon USA

I CERTIFY, ON HONOR, That I have minutely inspecte[d] previously to his enlistment, and that he was entirely sober judgment and belief, he is of lawful age; and that, in accepting of an able-bodied soldier, I have strictly observed the Regulat[ions] This soldier has Blue eyes, Dark hair, 5 8, inches high.

Regiment of

(A. G. O. No. 74 & 75.)

Mustered into the service of the United States, in Company ..., Regiment of ..., the ... day of ..., 186 , at ...

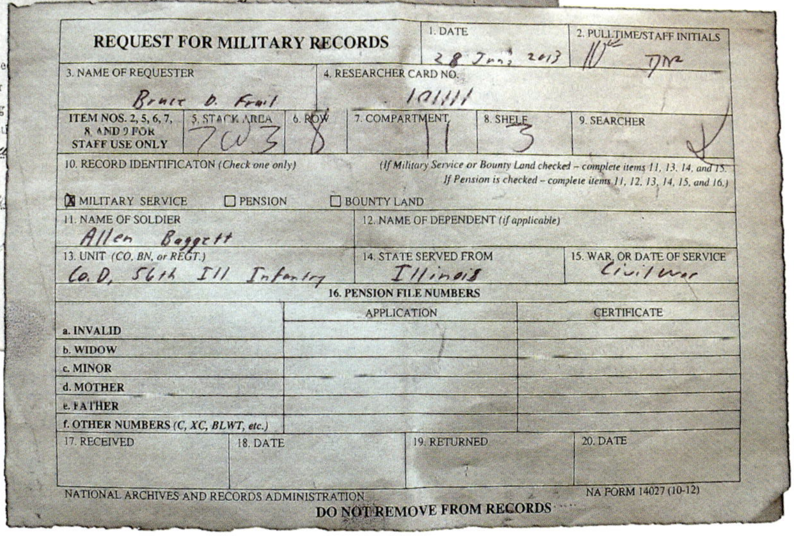

REQUEST FOR MILITARY RECORDS

1. DATE 28 June 2013 — 2. FULLTIME/STAFF INITIALS

3. NAME OF REQUESTER Bruce D. Frail — 4. RESEARCHER CARD NO. 101111

ITEM NOS. 2, 5, 6, 7, 8, AND 9 FOR STAFF USE ONLY — 5. STACK AREA 7W3 — 6. ROW 8 — 7. COMPARTMENT 11 — 8. SHELF 3 — 9. SEARCHER

10. RECORD IDENTIFICATON (Check one only) (If Military Service or Bounty Land checked – complete items 11, 13, 14, and 15. If Pension is checked – complete items 11, 12, 13, 14, 15, and 16.)

☒ MILITARY SERVICE ☐ PENSION ☐ BOUNTY LAND

11. NAME OF SOLDIER Allen Baggett — 12. NAME OF DEPENDENT (if applicable)

13. UNIT (CO, BN, or REGT.) Co. D, 56th Ill Infantry — 14. STATE SERVED FROM Illinois — 15. WAR, OR DATE OF SERVICE Civil War

16. PENSION FILE NUMBERS

	APPLICATION	CERTIFICATE
a. INVALID		
b. WIDOW		
c. MINOR		
d. MOTHER		
e. FATHER		
f. OTHER NUMBERS (C, XC, BLWT, etc.)		

17. RECEIVED — 18. DATE — 19. RETURNED — 20. DATE

NATIONAL ARCHIVES AND RECORDS ADMINISTRATION — NA FORM 14027 (10-12)

DO NOT REMOVE FROM RECORDS

Request for Military Records form filled out in the library by the researcher. Includes precise location of records in the library, derived from onsite indexes.

Online CMSR Records

The majority of Compiled Military Service Records have not been digitized, still requiring the researcher to retrieve copies from the National Archives. It is difficult to know the total number of CMSRs in the National Archives, though we do know more than 1.2 million are scanned.

In 1994, NARA began a project in conjunction with National Park Service's *Civil War Soldiers and Sailors System* (CWSS) to index and microfilm all CMSRs. This was a massive effort retrieving and organizing regimental records from storage.

A primary goal of the project was to publish the database online. This included Union and Confederate units and soldiers; organized and indexed by soldiers, sailors, regiments, battles, and other subjects. Over 6,000 *Battle Units* (usually a regiment) were indexed, Union and Confederate. The current index database lists over six million soldiers, sailors, and marines.

Unfortunately, this index database includes little information about a veteran, just his basic information: Battle Unit Name, Side (Union or Confederate), Company, Rank In, Rank Out, and Alternate name. Yet it is a start for a new search.

> **Research Tip**: The National Park Service *Civil War Soldiers and Sailors System (CWSS)* is an excellent place to start research.
> **https://www.nps.gov/civilwar/soldiers-and-sailors-database.htm**

Most of the indexed data was retrieved from original CMSRs. Millions of individual records have not been microfilmed, which is crucial to online CMSR access. Microfilm is basically photographic technology and has been used for over 100 years to preserve documents. Like a photograph, microfilm needs to be digitally scanned to allow computer or online viewing.

Who Was Digitized?

Still, many CMSRs were microfilmed and subsequently scanned. The CMSR folders were organized by state, and each state was microfilmed.

CMSR Folders and Record Scans

Surprising to most researchers, many CMSR folders including the records inside were scanned. They can be found on Fold3 in state and regiment order. It must be noted however, that some scanned folders only contain one record while other folders may have dozens of records. These folders number over 1.2 million. Note: Folders are variously referred to as *files*, *jacket-envelopes*, and of course *folders*.

CMSR Index Only

Some state collections only include an *index* record for each folder. This gives little information except that it confirms a soldier's unit designation. These indexes may have been a preliminary step before the complete folder was scanned, as some states have both complete folders and indexes available. It appears that the scanning project ceased, maybe from budget cuts, in about 2016.

Digitized Records by State

The following table lists the number of scanned and indexed folders for each state. (Data from each state entry in Fold3, June 2021)

CMSR Folders Scanned and Indexed						
State	Scanned	Indexed		State	Scanned	Indexed
Alabama	3,549	164,906		Missouri	309,183	196,848
Arkansas	14,587	–		N. Carolina	6,227	–
California	23,061	45,507		New Jersey	–	97,060
Colorado	9,218	21,428		New Mexico	17,723	1,523
Wash. D.C.	12,537	8,470		Nebraska	5,548	–
Delaware	19,751	–		Nevada	5,548	–
Florida	2,247	–		N. H.	–	45,507
Georgia	214	–		New York	6,133	512,469
Illinois	–	371,142		Ohio	–	464,829
Indiana	–	277,640		Oregon	2,805	–
Iowa	–	93,389		Pennsylvania	–	509,597
Kansas	–	32,318		Rhode Island	–	26,534
Kentucky	172,397	–		S. Dakota	295	–
Louisiana	18,912	–		Tennessee	81,712	–
Maine	–	83,362		Texas	4,796	–
Maryland	92,051	–		Utah	96	–
Mass.	367,528	–		Vermont	49,213	43,649
Michigan	–	143,556		Virginia	73,935	–
Minnesota	–	31,009		Washington	–	1,523
Mississippi	1,029	–		Wisconsin	–	127,428

Access

CMSRs and indexes can be viewed online at Fold3. This is a subscription service, though many libraries offer free Fold3 access. Most NARA facilities also provide free access to Fold3.

1. Select SEARCH from the main menu at top.
2. In CONFLICT/WAR Select *Civil War (Union).*
3. In PUBLICATION TYPE, select *Service Records.*
4. PUBLICATION NAME will now show CMRSs for states, choose *See All.*
5. A pop-up window will list all the states.
6. Browse through the regiments listed and follow the records to an individual name.

> **Research Tip:** If only an index is available, the information on that index card can be referenced when ordering copies of the complete CMSR from the National Archives (NARA).

U.S. Colored Troops Online

One important, and nearly complete, subset of the CMSRs was digitized—the U.S. Colored Troops (U.S.C.T.) This force was comprised of 138 numbered infantry regiments, 17 numbered artillery regiments (heavy and light), seven numbered cavalry regiments, and numerous independent brigades and companies in the Union army. a majority of the records were microfilmed and scanned. This was the first set of CMSR records released.

U.S.C.T. Regiments – Digitized CMSR		
Infantry	**Artillery**	**Cavalry**
1st -138th	1st - 15th Heavy	2nd - 7th
–	1st - 2nd Light	54th Mass (Col)

Access

The nearly complete set of U.S.C.T. folders can be viewed online at Fold3.com.

1. Select SEARCH from the main menu at top.
2. In CONFLICT/WAR Select *Civil War (Union).*
3. In PUBLICATION TYPE, select *African American Collection.*
4. In PUBLICATION NAME choose *See All.*
5. All African American related records will be listed (over 320,000).
6. Enter CMSR in the Filter bar where it says *People, Records, Places.*
7. This will filter the list to just the CMSR records.
8. Browse through the regiments listed and follow the records to an individual name.

Additional Sources

The CMSR microfilm were made available to Ancestry and FamilySearch. However, those services have not organized the data as simply as Fold3. While both services will point to a specific record from search, you then need to browse through a large file to see all relevant records. The records are in alphabetical order by last name. Check these services regularly for updates.

Ancestry

Ancestry includes the U.S.C.T. dataset (*U.S., Colored Troops Military Service Records, 1863-1865*) but not the larger dataset of U.S. infantry, artillery, and cavalry regiments. Access is by a familiar Ancestry-type form. Search a person's name and a list of all occurrences will be returned. Select a record and browse through the associated records.

https://www.ancestry.com/search/collections/1107/

Family Search

The FamilySearch collection contains only a portion of the U.S.C.T. It is limited to 1st through 5th Cavalry, 5th Massachusetts Cavalry, 6th Cavalry, 1st Infantry, 1st South Carolina Volunteers Company A, 1st Infantry, and 36th through 40th Infantry.

https://www.familysearch.org/search/collection/3518724

Miscellaneous Records

Some databases list miscellaneous CMSR cards that are not associated with a soldier's packet. Insufficient or contradictory information made it difficult to link these records to a particular soldier. These records were collected with the intent that they would eventually become part of the compiled service records.

CMSR Jacket Sample

The jacket/folder/envelope was a narrow folder with a single pocket to place 7" x 3" cards that were transcribed from original records. Larger, full sheet documents such as original enlistment papers, were folded to fit. The front of the jacket included a place to enter a unique tracking number of each document/record.

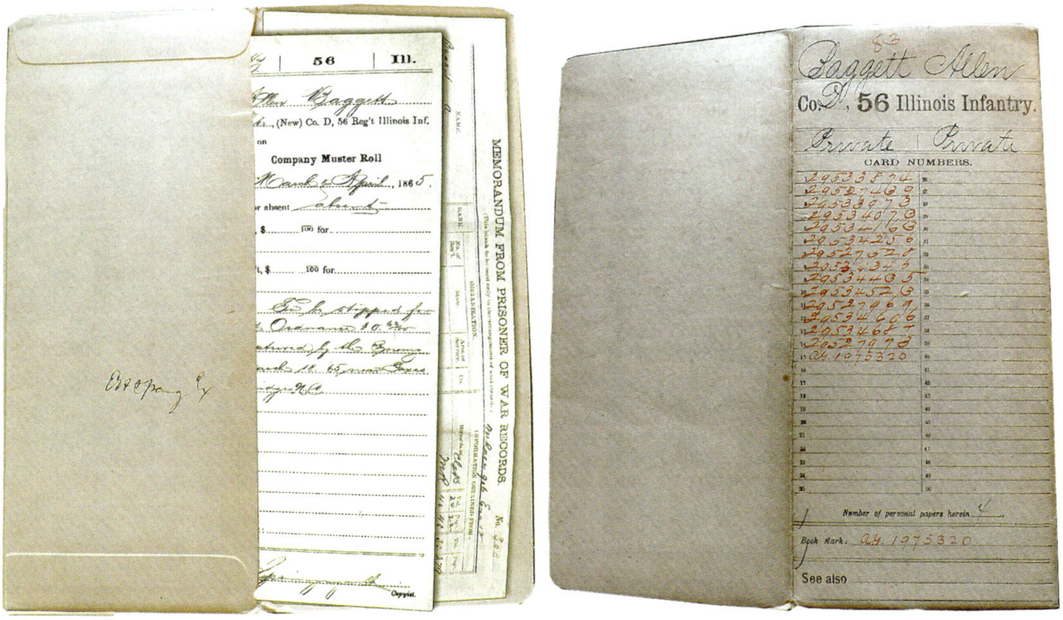

CMSR Jacket/Folder of Allen Baggett, Co. D, 56th Illinois Infantry.
Left image is inside of jacket with documents, right is front of jacket with list of documents inside.

Images of Allen Baggett's (a great-great-grandfather of the author) CMSR were obtained in 2013 through a service that visits the National Archives and makes digital copies of files. A CD was provided with 27 CMSR images and over 100 pension file images. Today, most services provide files through a download system.

The CMSR file contained enlistment papers, Muster Rolls which tracked various events including his capture by the enemy, an Adjutant General report verifying his service to the Pension Bureau, Prisoner of War Records, and jacket images.

Allen Baggett (1845 - 1941), Private, Co. D, 56th Illinois Infantry. Above Company Muster Roll includes *"Captured by enemy March 18 '65 near Foxes Bridge, N.C."* He finished the war in Libby Prison in Richmond, VA. (Photo c. 1941)

Pension Records

There was a direct connection between the G.A.R. and Union pension benefits. The G.A.R.'s political influence created a veteran-friendly Congress that passed most pension legislation brought to them, expanding eligibility and increasing payments. So many veterans were applying for a pension by the late 1880s that the Pension Bureau was overwhelmed processing them. To accommodate the processes and storage needs, the Pension Bureau built an enormous building, which at the time was the largest brick building in the world. The bureau's work produced records that give us a unique view into the lives of many Union veterans.

Clerks and pension files in the 1930s, and a 1902 view of the Pension Building. (Library of Congress)

Pension application and subsequent payment increase requests entailed more than filling out a simple form. In most cases medical examinations were needed, affidavits from people who knew the veteran's condition were required, and an endless exchange of letters all left a lengthy paper trail. Even a change of address notice to insure they received their check left evidence of a veteran's journey through life.

The ultimate treasure chest of veteran information is his pension file, which usually contains the original version of every correspondence. This author received his great-great-grandfather's pension file (digitized) which contained over 150 images of documents and correspondence. Even upon his death, documents continued to accumulate. There were claims for burial assistance, notification of suspension of pensions, and even a copy of his death certificate among the documents, which I did not have.

Various pension records are available from different sources, even genealogy web services. The most commonly found record is the Pension Index Card. It contains brief but very valuable information that may point you to a full pension file.

Pension Index Cards

Finding the veteran's pension index card is crucial to identifying a complete pension file. Luckily, these are available online. Several aspects of this card are important. First, it verifies that your veteran did apply for a pension. Second, his name and regiment match other records, so this is him. And third, the *Date of Filing*, *Application No.* and *Certificate No.* are listed, along with the *State From Which Filed*. Finally, while it may not be under *Remarks* for every file, in the Allen Baggett example C2575971 is the actual number written on the outside of the physical file. You wouldn't have known this until you saw the file, but an archivist in the National Archives would recognize it as a folder number.

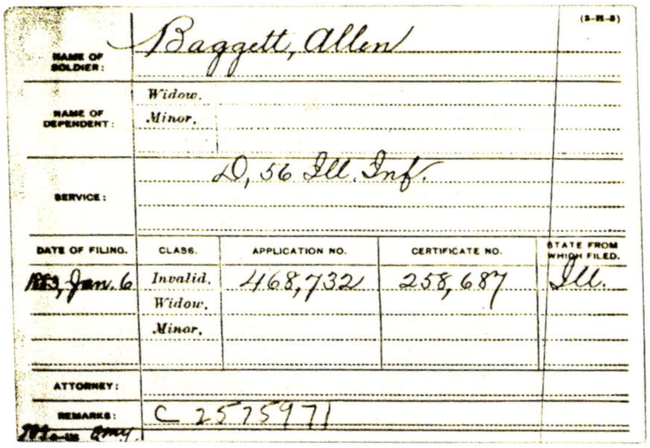
NAME OF SOLDIER: Baggett, Allen
NAME OF DEPENDENT: Widow. Minor.
SERVICE: D, 56 Ill. Inf.

DATE OF FILING.	CLASS.	APPLICATION NO.	CERTIFICATE NO.	STATE FROM WHICH FILED.
1883, Jan. 6	Invalid.	468,732	258,687	Ill.
	Widow.			
	Minor.			

ATTORNEY:
REMARKS: C 2575971

Record Access

Ancestry https://www.ancestry.com/search/collections/4654/

Fold3 https://www.fold3.com/title/57/civil-war-pensions-index/

Pension Payment Cards

Printed index cards were used by the Bureau of Pensions and Veterans Administration to record the payment of pensions to veterans, widows and other dependents. One was created for every veteran, widow, or minor who received a pension.

On the card front was the veteran's or widow's name, certificate number, his unit, disability for which pensioned, pension increases, and other details. The reverse side is the payment detail.

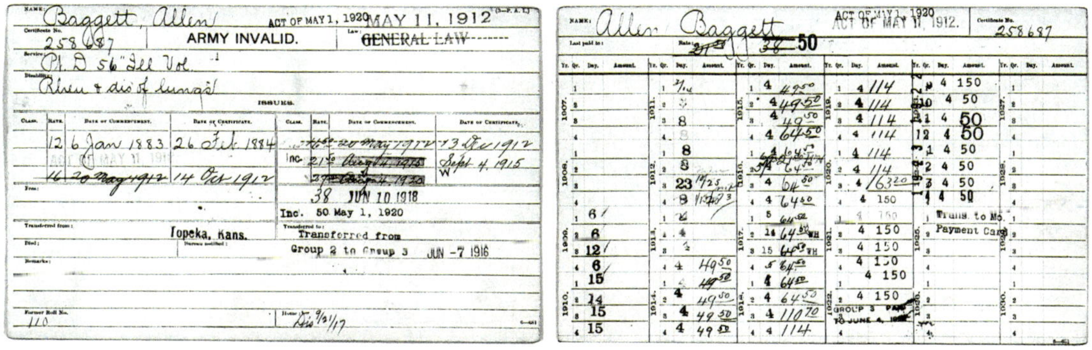
Baggett, Allen — ACT OF MAY 1, 1920 — MAY 11, 1912
258687 — ARMY INVALID. — GENERAL LAW
Co. D 56 Ill Vol
Topeka, Kans.
Inc. 50 May 1, 1920
Transferred from Group 2 to Group 3 JUN -7 1916
Allen Baggett — ACT OF MAY 11, 1912 — 258687

A Pension Payment Card

These 5" x 8" cards replaced leather bound payment ledgers beginning in 1907; the above record was begun in 1909. Still, the front recorded that his initial pension began February 26, 1884 at the rate of $12.00 per month. Pensions were initially paid quarterly, hence the *Amount* listed until 1922 was a quarterly payment. Beginning in 1923, payments were issued monthly, as his amount now reflects the monthly payment of $50.00, which included a raise.

A thorough explanation of Pension Payment Cards was published by Claire Prechtel-Kluskens, a senior reference and projects archivist at NARA of the National Archives, in *NGS Magazine*. Her blog introduces and includes a link to the 4-page article:

https://twelvekey.com/2015/05/30/pension-payment-cards-1907-1933/

NARA created 2,539 microfilm rolls of pension cards in the 1970s. Subsequently, Archives.org digitized those rolls which are now available on FamilySearch.

Records Access

FamilySearch provides a searchable database of Pension Payment Card images:

FamilySearch https://www.familysearch.org/search/collection/1832324

Another way to view the records is more difficult, but useful as well. It allows you to browse digitized microfilm by name ranges.

FamilySearch https://www.familysearch.org/wiki/en/Veterans_Administration_Pension_Payment_Cards,_1907%E2%80%931933

> **Research Tip**: Pension Payment Cards are also available for widows and dependents who received pensions.

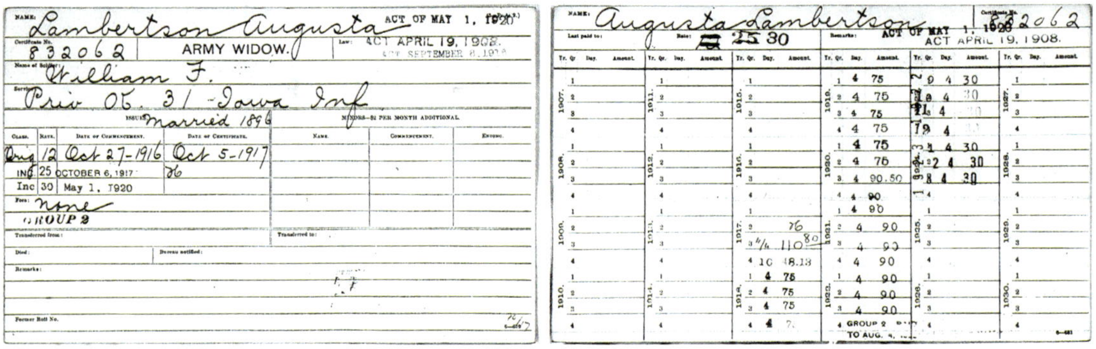
NAME: Lambertson Augusta ACT OF MAY 1, 1920
Certificate No. 832062 ARMY WIDOW. ACT APRIL 19, 1908. ACT SEPTEMBER 8, 1916
Name of Soldier: William F.
Service: Priv. Co. 31 - Iowa Inf
Married 1896
Aug 12 Oct 27-1916 Oct 5-1917
Inc 25 OCTOBER 6, 1917
Inc 30 May 1, 1920
Fees: none
GROUP 2

NAME: Augusta Lambertson Certificate No. 832062
$25 30 ACT OF MAY 1, 1920 ACT APRIL 19, 1908.

Widow Pension Payment Card

Upon death of a veteran who was receiving a pension, the pension could pass on to his widow and minor dependents. The requirements changed significantly over the years; the first recipients were only widows of soldiers killed during the war, though eventually all widows were eligible. Minor children were generally eligible up to 16 years of age.

Augusta Lambertson started receiving a widow's pension in October 1916, one year after her Civil War veteran husband died, who had been receiving a pension since 1907. Her card shows she received $12 a month in 1916, with increases to $25 in 1917 and $30 in 1920. William F. Lamberston died in October of 1915; there is no indication why there was a one-year lapse of her receiving the pension. Possibly she delayed applying or Pension Bureau paperwork caused the delayed.

Pension Files

You can help researchers at NARA, and increase the probability of finding pension records by providing all the veteran data you have. This author did not realize that *C2575971* on the index card was the actual folder number, until images were received from a hired researcher.

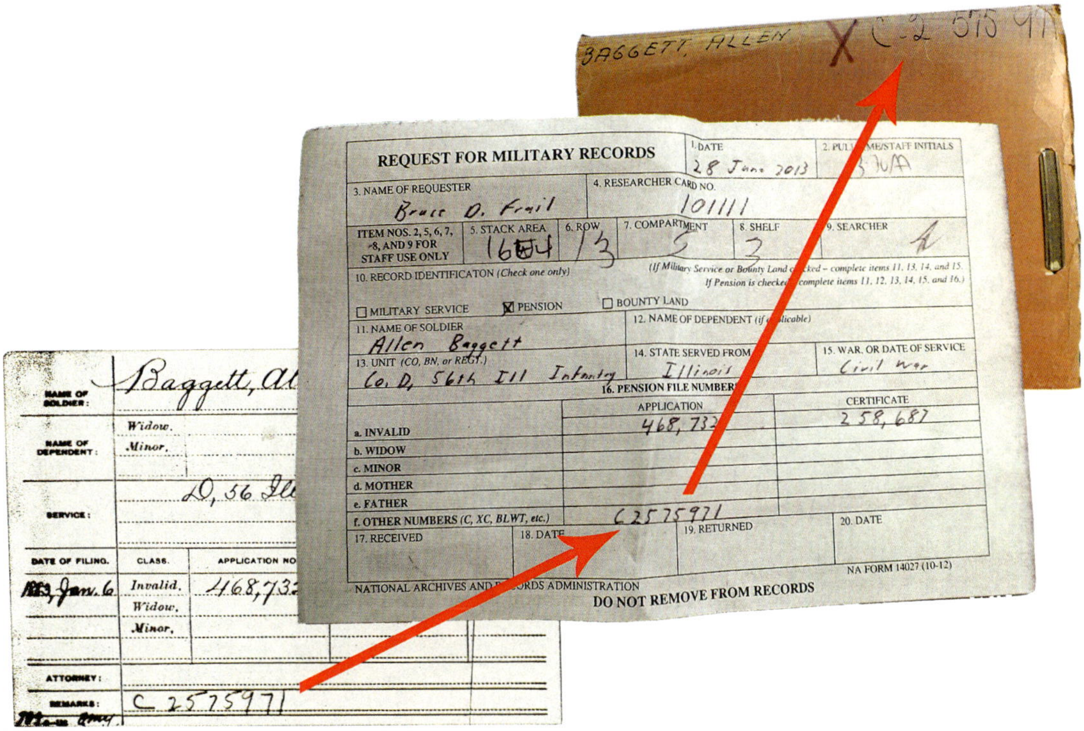

BAGGETT, ALLEN X C-2 575 971

REQUEST FOR MILITARY RECORDS
1. DATE 28 June 2013
2. PULL TIME/STAFF INITIALS
3. NAME OF REQUESTER Bruce D. Frail
4. RESEARCHER CARD NO. 101111
ITEM NOS. 2, 5, 6, 7, 8, AND 9 FOR STAFF USE ONLY
5. STACK AREA 16E4
6. ROW 13
7. COMPARTMENT 5
8. SHELF 3
9. SEARCHER
10. RECORD IDENTIFICATON (Check one only)
(If Military Service or Bounty Land checked – complete items 11, 13, 14, and 15. If Pension is checked complete items 11, 12, 13, 14, 15, and 16.)
☐ MILITARY SERVICE ☒ PENSION ☐ BOUNTY LAND
11. NAME OF SOLDIER Allen Baggett
12. NAME OF DEPENDENT (if applicable)
13. UNIT (CO, BN, or REGT.) Co. D, 56th Ill Infantry
14. STATE SERVED FROM Illinois
15. WAR, OR DATE OF SERVICE Civil War
16. PENSION FILE NUMBERS

	APPLICATION	CERTIFICATE
a. INVALID	468,732	258,687
b. WIDOW		
c. MINOR		
d. MOTHER		
e. FATHER		
f. OTHER NUMBERS (C, XC, BLWT, etc.)	C2575971	

17. RECEIVED 18. DATE 19. RETURNED 20. DATE
NATIONAL ARCHIVES AND RECORDS ADMINISTRATION
NA FORM 14027 (10-12)
DO NOT REMOVE FROM RECORDS

NAME OF SOLDIER: Baggett, Al
NAME OF DEPENDENT: Widow. Minor.
SERVICE: D, 56 Ill
DATE OF FILING. 1883, Jan. 6 CLASS. Invalid. Widow. Minor. APPLICATION NO. 468,732
ATTORNEY:
REMARKS: C 2575971

Example of the Pension Index Card directing a researcher towards the complete pension folder.

The pension file can reveal surprising family history. Allen Baggett's file divulged a wife previously unknown to his descendants. His first wife died in 1913; she is well documented and is a grandmother to many, including the author. A subsequent wife, Frances Baggett, stated in a 1924 *Attendant's Affidavit in Invalid Pension Claim* that she married Baggett on May 26, 1920.

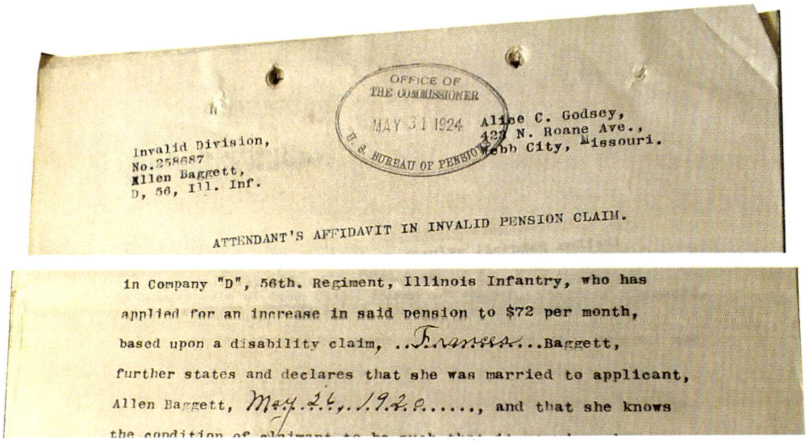

OFFICE OF THE COMMISSIONER MAY 31 1924 U. S. BUREAU OF PENSIONS

Invalid Division,
No.258687
Allen Baggett,
D, 56, Ill. Inf.

Alice C. Godsey,
429 N. Roane Ave.,
Webb City, Missouri.

ATTENDANT'S AFFIDAVIT IN INVALID PENSION CLAIM.

in Company "D", 56th. Regiment, Illinois Infantry, who has applied for an increase in said pension to $72 per month, based upon a disability claim, ..Frances...Baggett, further states and declares that she was married to applicant, Allen Baggett, May 26, 1920......, and that she knows

A veteran's attendant needed to attest to a disability during a new claim. It could be a nurse, friend, or relative. In Allen Baggett's case, his wife. They were married for 15 years until she died in 1935. They lived in a state away from the extended family and no stories or evidence of her were known to current descendants until this document appeared.

Affidavits like the above are typical of documents found in pension files. Statements from doctors and attendants were frequently required for pension updates. Other types of documents that may be found include:

- Initial pension claims and Declaration for Pension Increases
- Medical examinations
- Adjutant General's (War Department) confirmation of military service
- Request for Burial Flag or Claim for Burial Allowance (Expenses)
- Death Certificate and many others

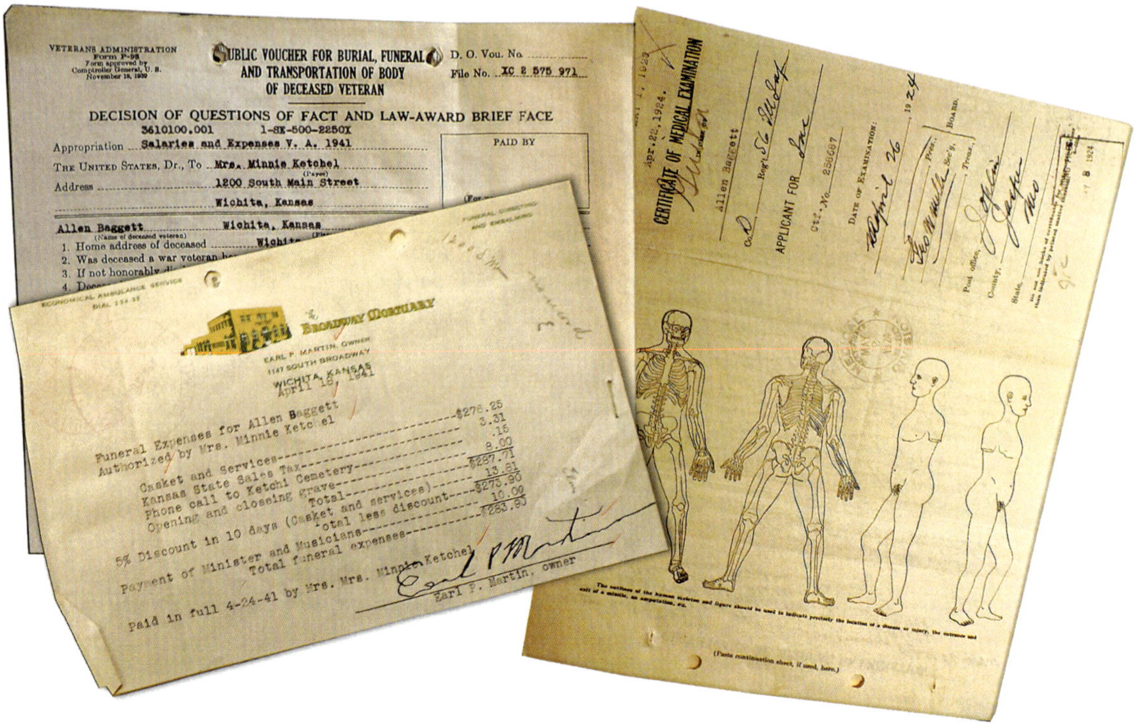

Typical Records from a Pension File: Medical Exam Certificate, Voucher for Death, Expenses, receipts.

Record Access

1. Order Copies
2. Hire A Service (see page 82).
3. Visit the National Archives and make copies yourself (see page 82).

Order Copies – Begin with the NARA web page titled *Requesting Copies of Older Military Service Records.*

https://www.archives.gov/veterans/military-service-records/pre-ww-1-records

Look for *Military Pension/Bounty-Land Warrant Applications*; pension file ordering is different than for CMSR files. You are given two choices for ordering, one via an on-line request, the other is submitting a paper form (NATF 85) through the mail. Note: If you choose to order on-line, you will be required to create an account with National Archives, which is simple and free.

Missing Records

Pension records, particularly in the 20th century, were occasionally moved to new locations or jurisdictions. You may have to submit requests to multiple sources. Start with NARA using the steps outlined here and in their guidelines.

X and XC Pension Records

Some Pension Index Cards list a folder/file number in the remarks section, as shown on Allen Baggett's card on page 91. Some have no number, others a 'C' or 'XC.' The X series designations (and others) may indicate where the full pension file is located. See Appendix B for a discussion on file locations.

Additional Records and Sources

Retrieving copies of Civil War veteran records from the National Archives can deliver unexpected rewards. Still, there are other sources that provide clues to their lives.

Special Ancestry Search

Ancestry has many collections that can be searched one at a time; several were described earlier. They also created a special search that looks at multiple collections during one search, and it may uncover Confederate records.

Ancestry https://www.ancestry.com/cs/civilwarrecords

National Parks Service (NPS)

The National Parks Service (NPS) built the *Soldiers and Sailors* database, which is useful but has its quirks. It was created by the NPS with two primary partners, the Genealogical Society of Utah (GSU) and the Federation of Genealogical Societies (FGS) as part of the Civil War Sesquicentennial commemoration beginning in 2011.

Every soldier who participated in the war, from either the Union or the Confederacy, *should* have a record on this website; however casual searches have not found some known veterans. The source of this data was the *General Index Cards* of the *Compiled Military Service Records,* at the National Archives. Other records linked to the names from the Index Cards were also consulted. The same database is available from FamilySearch and Ancestry.

NPS https://www.nps.gov/civilwar/soldiers-and-sailors-database.htm
FamilySearch https://www.familysearch.org/search/collection/1910717
Ancestry https://www.ancestry.com/search/collections/1138/

National Cemetery Administration (Dept of VA)

The National Cemetery Administration (NCA), a function of the Veterans Affairs Department, created a *Gravesite Locater* system that contains all veteran cemeteries and their graves, whether national or state. It is available online and most of these cemeteries offer access via a kiosk with a touch screen monitor. Ancestry also offers access to the database.

NCA https://gravelocator.cem.va.gov/
Ancestry https://www.ancestry.com/search/collections/8750/

Ft. Scott National Cemetery includes a Gravesite Locator.

Census Records

Census records are not the first place most researchers consider when looking for Civil War veteran information. Yet, they can lend some valuable information to a veteran's history.

The number of unexpected pension applications increased dramatically in the late 1800s, overwhelming the Pension Bureau. This prompted the government to add veteran questions to the census hoping this would predict future application demand. More than just another box to check, a separate detailed schedule was added to the standard 1890 census.

1890 Population Census

The most famous aspect of the 1890 census is that most of it did not survive a disastrous 1921 fire in a Commerce Department building. Some records did survive, but they offer a sketchy version of the census. A unique feature of the 1890 census was the format: a separate page was used for each family or dwelling, with the data organized in columns for an individual. This census did provide a line to identify veterans.

Question Two on the general population schedules inquired about the individual:

> *"Whether a soldier, sailor, or marine during the civil war (United States or Confederate) or widow of such person."*

Enumerators were instructed to write "Sol" for soldier, "Sail" for sailor, and "Ma" for marine, with "U.S." or "Conf." in parentheses for every veteran, For example, Sol (U.S.) or Sail (Conf.). The letter "W" was added to these designations if the enumerated was a widow. If one of these applied, the *Veteran Schedule* was then completed for that person.

> **Research Tip**: The surviving 1890 census schedules are available for a few scattered townships or counties in Alabama, District of Columbia, Georgia, Illinois, New Jersey, New York, North Carolina, Ohio, South Dakota, Texas.

The 1890 population census is available on Ancestry.

Ancestry **https://www.ancestry.com/search/collections/5445/**

1890 Veteran Schedule

The Pension Bureau asked the Interior Department to request detailed veteran information during the 1890 enumeration. This March 1, 1889 authorization included:

> *That said Superintendent shall, under the authority Special inquiries of the Secretary of the Interior, cause to be taken on a special schedule of inquiry . . . the names, organizations, and length of service of those who had served in the Army, Navy, or Marine Corps of the United States in the war of the rebellion, and who are survivors at the time of said inquiry, and the Survivors of the war widows of soldiers, sailors or marines.*

This is commonly mistaken as a separate census, while it was actually an additional schedule used while taking the general population census. If the main census included a veteran, details of the veteran's service was recorded on the new schedule titled: *Special Schedule—Surviving Soldiers, Sailor, and Marines, and Widows.*

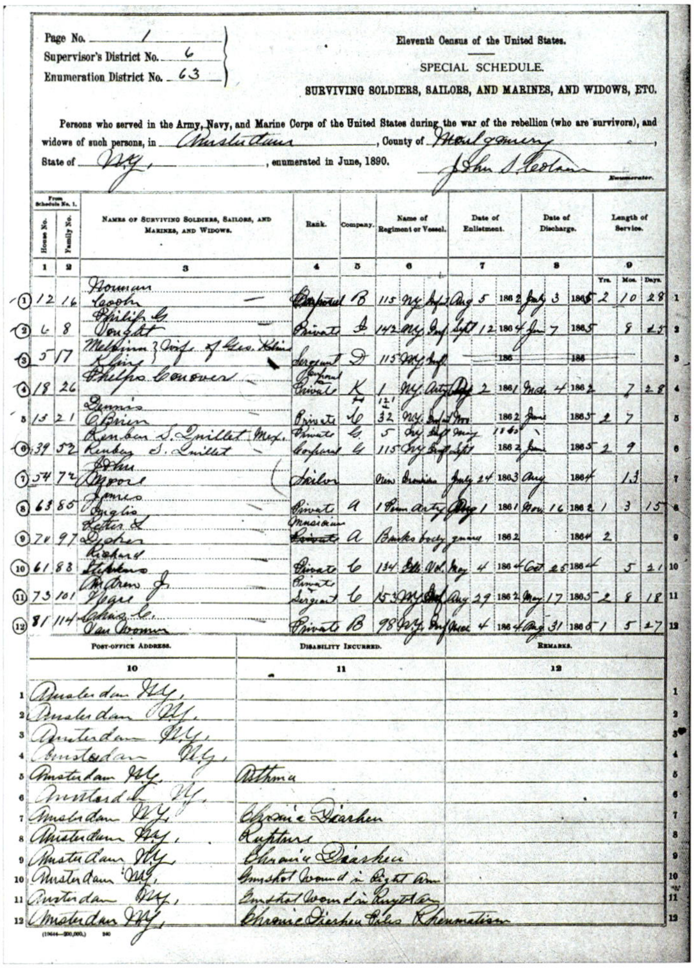
Page No. 1
Supervisor's District No. 6
Enumeration District No. 63

Eleventh Census of the United States.
SPECIAL SCHEDULE.
SURVIVING SOLDIERS, SAILORS, AND MARINES, AND WIDOWS, ETC.

Persons who served in the Army, Navy, and Marine Corps of the United States during the war of the rebellion (who are survivors), and widows of such persons, in Amsterdam, County of Montgomery, State of N.Y., enumerated in June, 1890.

Names of Surviving Soldiers, Sailors, and Marines, and Widows. | Rank. | Company. | Name of Regiment or Vessel. | Date of Enlistment. | Date of Discharge. | Length of Service.

Post-office Address. | Disability Incurred. | Remarks.

Amsterdam N.Y.
Asthma
Chronic Diarrhea
Rupture
Chronic Diarrhea
Gunshot Wound in Right Arm
Chronic Diarrhea Piles Rheumatism

Schedule includes space for:

- Names of surviving soldiers, sailors, marines, and widows
- Rank
- Regiment or Ship
- Date of enlistment
- Date of discharge
- Length of service
- Post office address
- Disability incurred
- Remarks

Record Access

Like the 1890 general census, most of the veteran schedules were victims of the 1921 fire, with all of Alabama through Kansas, and much of Kentucky, destroyed. Parts of Kentucky through Wyoming are available.

All veterans or their widows in the 1890 census, even from states that survived the fire, may not be listed. An audit of the enumeration found many veterans were missed. Also, while this was intended to identify Union veterans (for pension planning), some Confederate soldiers are found. The available1890 Veteran Census schedules can be searched at Ancestry or FamilySearch.

Ancestry	https://www.ancestry.com/search/collections/8667/
FamilySearch	https://www.familysearch.org/search/collection/1877095

Headstones of Deceased Veterans

A collection of records called *Headstones Provided for Deceased Union Civil War Veterans, 1861-1904* are available online.

After the Civil War, deceased Union veterans were eligible for burial in a National Cemetery, which included a headstone. An act of Congress on February 3, 1879, extended the privilege of government-provided gravestones to soldiers buried in private cemeteries. This act provided:

> *. . . That the Secretary of War is hereby authorized to erect headstones over the graves of soldiers who served in the Regular or Volunteer Army of the United States during the war for the Union, and who have been buried in private village or city cemeteries, in the same manner as provided by the law of March third, eighteen hundred and seventy-three, for those interred in national military cemeteries . . .*[2]

Headstone and record (below) for Cornelius Decker, 1847- 1879

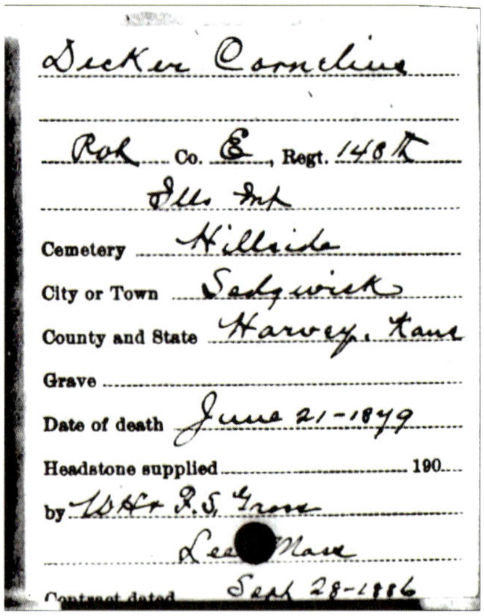

Decker Cornelius

Pvt. Co. E, Regt. 148th

Ills. Inf

Cemetery Hillside

City or Town Sedgwick

County and State Harvey, Kans

Grave

Date of death June 21-1879

Headstone supplied 190

by [illegible]

[illegible]

Contract dated Sept 28-1886

This popular act set off a flurry of headstone applications for soldiers buried in private and public cemeteries. These records give us one more piece of interesting history.

Over 166,000 cards were created by the Office of the Quartermaster General that documented the applications. They were subsequently microfilmed, eventually digitized, and provided to online services.

Many times the headstone was ordered and placed years after burial, including numerous cases where a G.A.R. post canvassed local cemeteries and ordered blocks of headstones for their deceased comrades.

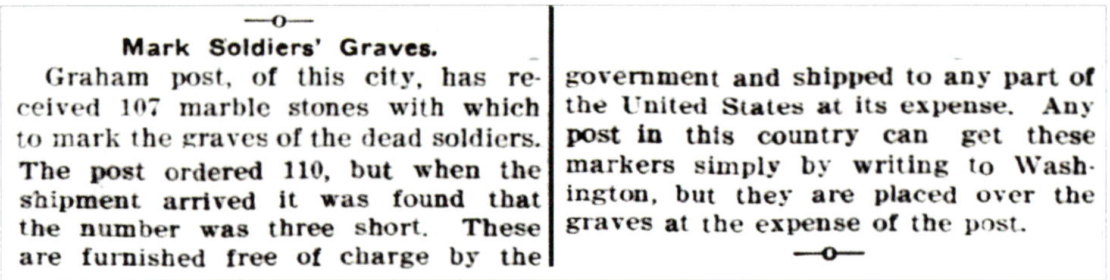

Mark Soldiers' Graves.

Graham post, of this city, has received 107 marble stones with which to mark the graves of the dead soldiers. The post ordered 110, but when the shipment arrived it was found that the number was three short. These are furnished free of charge by the government and shipped to any part of the United States at its expense. Any post in this country can get these markers simply by writing to Washington, but they are placed over the graves at the expense of the post.

The Rock Island Argus (Rock Island, Illinois) June 17, 1905.

Record Access

Ancestry	https://www.ancestry.com/search/collections/1195/
FamilySearch	https://www.familysearch.org/search/collection/1913388

7 State and Local Records

Regimental Histories • Adjutant General Reports • Bounty Records • Personal War Sketches

Since most regiments were organized by state, their recruitment and history records were collected and published by the respective states. These records can be just as valuable as federal records, though they may be harder to find.

State regiments were chronicled in four kinds of documents which are typically found in state archives and libraries:

- Regimental Histories
- Adjutant General Reports
- Bounty Records
- Personal War Sketches

While each document covers history within a specific state, many times they can be found in archives across the country. The records created by individual states vary widely. Eastern states are more likely to have preserved their records, as they held a larger number of veterans who pushed for such preservation.

Only the Personal War Sketches strictly cover G.A.R. members, though personal diaries and accounts of other soldiers have been published. The remaining three books or documents address unit and soldier activity during and at the end of the war, before the G.A.R. was formed. All can add detail and personal insight to the history of a veteran.

Regimental Histories

Regimental histories are a fascinating compilation of regiment organization, campaigns, battles, and references to individual soldiers. While these are not specific to G.A.R. members, they offer details and sometimes personal insight into the life of the soldier before he became a member. Many times these were written by G.A.R. members and then handed to professional writers to finalize.

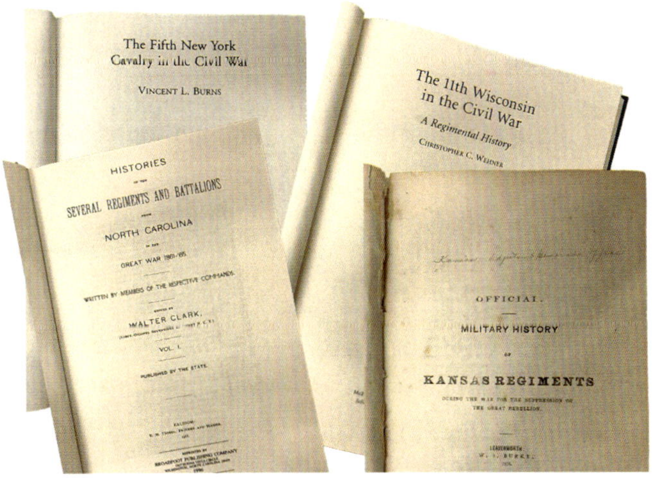

Each regimental history may cover detail differently as there was no standard format. Timothy J. Orr, author and Assistant Professor of History at Old Dominion University, aptly described a typical process in his excellent essay *Regimental Histories*:

> Returning Union soldiers compiled their regiments' official papers and handed them to accomplished authors, typically men who had served in their regiment from its official muster-in to its final muster-out. Veteran-authors sifted through hundreds of pages of official correspondence, interweaving the play-by-play of Civil War military campaigns with the various untold tales, personal letters, diary entries, and anecdotes told by the survivors.[1]

If you recall, a regiment was approximately 600-1,000 soldiers, usually recruited from a local area. Regimental histories are most likely found in states with a large number of regiments.

Approximate Number of Union Regiments from States and Territories*

State	Regiments	State	Regiments	State	Regiments
Alabama	2 (5)	Maryland	30	Pennsylvania	260+
Arkansas	11 (7)	Massachusetts	101 (3)	Tennessee	38 (2)
California	13	Michigan	79 (1)	Texas	3
Connecticut	37	Minnesota	22	Vermont	22
Delaware	13	Mississippi	1 (8)	Virginia	132+
Florida	4	Missouri	76 (5)	West Virginia	29
Georgia	1	Nevada	2	Wisconsin	58
Illinois	199	New Hampshire	21	Colorado Terr.	6
Indiana	167	New Jersey	48	Dakota Terr.	1
Iowa	60	New York	162	Nebraska Terr.	4
Kansas	47	North Carolina	8	New Mexico Terr.	5
Kentucky	69	Ohio	220+	Washington Terr.	1
Louisiana		Oregon	2		

* Above regiments include infantry, cavalry, & artillery. Numbers in parentheses were colored state regiments converted to Federal regiments. Regiment count is subject to interpretation, as some were mustered for only 100 days, some never completed organization. In addition, there were many detached or independent infantry and cavalry companies, and artillery batteries. Data was compiled from various online databases and state agencies, and do not include militia or home guards.

Many regimental histories are found in libraries, historical societies, and online. There are no complete Federal resources of regimental histories, though the Library of Congress and National Archives may have some in their catalogs. Regimental history titles varied, making identification difficult. Here are examples, all which can be called *Regimental Histories*:

- *A History of the Ninth Regiment Illinois Volunteer Infantry*
- *The Tenth Regiment Massachusetts Volunteer Infantry 1861 - 1864*
- *History of the 101st Regiment, Pennsylvania Veteran Volunteer Infantry 1861-1865*
- *Campaigns and Battles of the Sixteenth Regiment, Tennessee Volunteers*

Example of bound Regimental Histories.

Regimental History Access

Internet Archive

This is one of the best archives of old books, journals, and histories on any subject. In the search box, enter *[State] Regimental History* to start, with *State* being the one you are interested in.

https://archive.org/details/texts

HathiTrust

Many regimental histories are archived here. Search for *Regimental History.*

https://www.hathitrust.org/

Higginson Book Company, LLC

Higginson offers over 600 reprints of Civil War rosters and regimental histories. Use this as an index to search for a title in a library or historical society. If the book is not in your local library and it is from your state, suggest the library acquire a copy.

https://higginsonbooks.com/collections/civil-war

John Banks Civil War Blog

This page links to over 200 individual regimental histories on Internet Archive.

http://john-banks.blogspot.com/p/regimental-histories.html

Library of Congress (LOC)

Narrow your search to *Books/Printed Material* in the drop-down box and then enter *Regimental History* in the search box. If a book is located, it may not be available online; however, note the title and search in other sources.

https://loc.gov/

Libraries, Historical Societies, State Archives

Search these organizations using your *state* and *regiment history.* If a book is not in your library's holdings, use the Inter-Library Loan program to find an available copy.

General Internet Search

Use a search engine to conduct a general search using a variety of combinations of your *state* and *regiment history*, and a specific regiment name/number.

War of Rebellion: A Compilation of the Official Records . . .

The U.S. Military Academy states "T*he War of the Rebellion: a Compilation of the Official Records of the Union and Confederate Armies*, is the principal body of official orders and correspondence that document the course of military operations in all theaters of the American Civil War."[3]

This 130-Volume set contains the Official Records (O.R.) of regimental operations, creating a raw compilation of regimental histories. It is available in many libraries and on the internet. Cornell University Library created a research-friendly webpage that includes each volume number, the states included, and the time period covered. Use the "Search in this text" feature to find a specific regiment. This extensive compilation is also available on HathiTrust.

http://collections.library.cornell.edu/moa_new/waro.html
https://catalog.hathitrust.org/Record/000625514

Adjutant General Reports

An Adjutant General (AG) is a state's top military commander. In modern times, they generally are in charge of a state's National Guard. However, during the Civil War, state AGs shared command of state regiments with the Federal government, which also had an Adjutant General. A 2013 U.S. Army report titled *A Short History of the U.S. Army Adjutant General's Corps* described the relationships as:

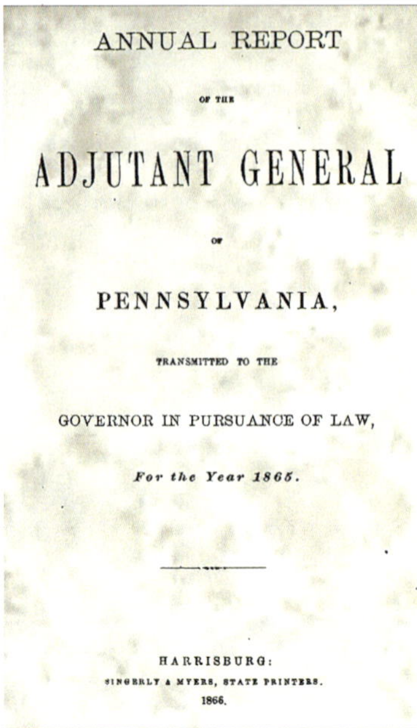
ANNUAL REPORT

OF THE

ADJUTANT GENERAL

OF

PENNSYLVANIA,

TRANSMITTED TO THE

GOVERNOR IN PURSUANCE OF LAW,

For the Year 1865.

HARRISBURG:

SINGERLY & MYERS, STATE PRINTERS.

1866.

> During the War Between the States the [Federal] Adjutant General's Department shared the task of mobilizing troops with the Adjutant General from each of the respective states. State militias were still under the authority of the state governors. Under this system, the Federal Government issued calls for troops, but the Governors and their Adjutant Generals were responsible for recruiting and initially equipping the troops until they were mustered into service. State Adjutant Generals kept records of each citizen who served in Civil War armies.[2]

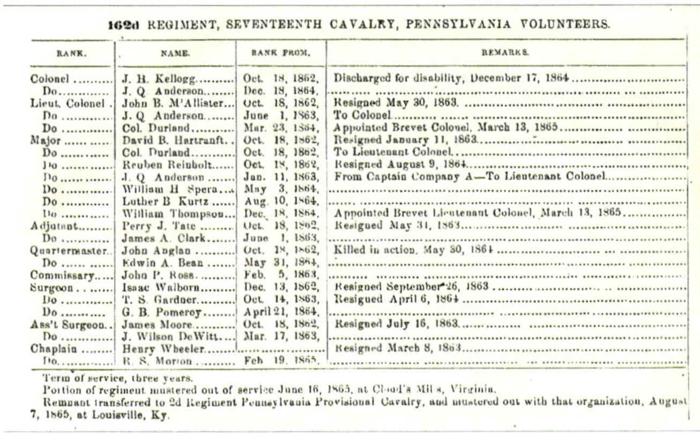
162d REGIMENT, SEVENTEENTH CAVALRY, PENNSYLVANIA VOLUNTEERS.

RANK.	NAME.	RANK FROM.	REMARKS.
Colonel	J. H. Kellogg	Oct. 18, 1862,	Discharged for disability, December 17, 1864
Do	J. Q Anderson	Dec. 18, 1864,	
Lieut. Colonel	John B. M'Allister	Oct. 18, 1862,	Resigned May 30, 1863.
Do	J. Q Anderson	June 1, 1863,	To Colonel
Do	Col. Durland	Mar. 23, 1864,	Appointed Brevet Colonel, March 13, 1865
Major	David B. Hartranft	Oct. 18, 1862,	Resigned January 11, 1863
Do	Col. Durland	Oct. 18, 1862,	To Lieutenant Colonel
Do	Reuben Reinbolt	Oct. 18, 1862,	Resigned August 9, 1864
Do	J. Q Anderson	Jan. 11, 1863,	From Captain Company A—To Lieutenant Colonel
Do	William H Spera	May 3, 1864,	
Do	Luther B Kurtz	Aug. 10, 1864,	
Do	William Thompson	Dec. 18, 1864,	Appointed Brevet Lieutenant Colonel, March 13, 1865
Adjutant	Perry J. Tate	Oct. 18, 1862,	Resigned May 31, 1863
Do	James A. Clark	June 1, 1863,	
Quartermaster	John Anglan	Oct. 18, 1862,	Killed in action, May 30, 1864
Do	Edwin A. Bean	May 31, 1864,	
Commissary	John P. Ross	Feb. 5, 1863,	
Surgeon	Isaac Walborn	Dec. 13, 1862,	Resigned September 26, 1863
Do	T. S. Gardner	Oct. 14, 1863,	Resigned April 6, 1864
Do	G. B. Pomeroy	April 21, 1864,	
Ass't Surgeon	James Moore	Oct. 18, 1862,	Resigned July 16, 1863.
Do	J. Wilson DeWitt	Mar. 17, 1863,	
Chaplain	Henry Wheeler		Resigned March 8, 1863
Do	R. S. Morton	Feb 19, 1865,	

Term of service, three years.

Portion of regiment mustered out of service June 16, 1865, at Cloud's Mills, Virginia.

Remnant transferred to 2d Regiment Pennsylvania Provisional Cavalry, and mustered out with that organization, August 7, 1865, at Louisville, Ky.

ADJUTANT GENERAL. 233

1865 Pennsylvania Adjutant General Report lists soldiers in a regiment, along with useful details.

When the war ended, the regiment was returned to state command and discharged, but not before a thorough report was compiled by the state Adjutant General.

These reports hold similar information as some regimental histories, though with fewer personal stories. However, these reports may include campaigns reports and rosters of men such as this Pennsylvania report of 1865. *Adjutant General Reports* are almost always found in state archives and libraries, though some may be found in the National Archives, Smithsonian, and some online services.

Adjutant General Report Record Access

State Archives, Historical Societies, and Local Libraries

See Chapter Eight, State Resources, for listings.

Internet Archives

Includes an extensive collection of state AG Reports. Include the following search terms, to restrict returns to the Civil War era: *1865 Adjutant General Report.* Otherwise all reports through modern times will be returned. Also, use 1861, 1862, 1863, 1864 in separate searches. Including a state name will reduce returns further.

https://archive.org/details/texts

Bounty Records

To ensure enough troops were available to fight the Civil War, the Federal Government set soldier quotas for each state. These quotas were then assigned by the state to cities, towns, and townships based on population.

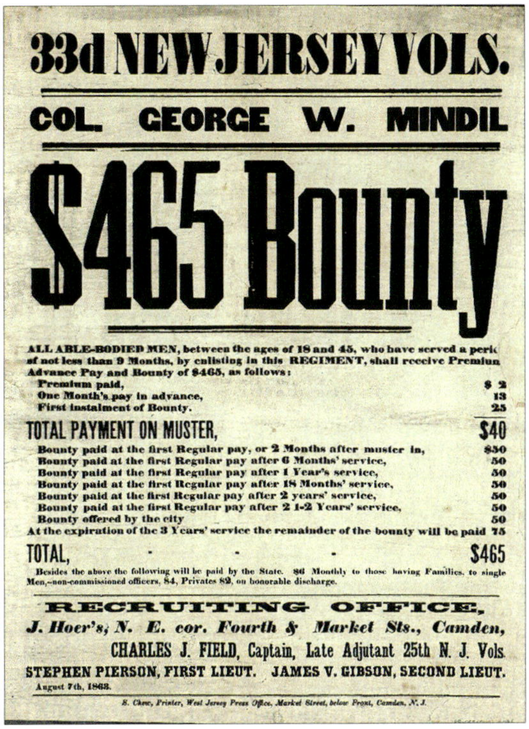

These quotas were satisfied in four manners.

- Instilling volunteers with patriotism.
- Enticing recruits with bounties or cash bonuses.
- Paying a substitute to take one's place.
- And finally, a conscription (draft) was introduced.

The volunteer method initially worked as Union overconfidence created a feeling that rebellious Confederates would melt away when confronted. Such was not the case.

Two of the options, bounty and draft, were fraught with inconsistencies from state to state, even town to town. In short, the bounty system became a mess, rife with fraud.

Creative posters flooded the streets with enticements to join the Army for any number of reasons—usually bounty money.

There were federal bounties, state bounties, county bounties, township bounties, and city bounties. Sometimes a town bounty competed with adjoining town and county bounties. The Fourth N.Y. Battery (at right) even offered a bounty from the Sixth (Congressional) District. These uncoordinated practices led some recruits to shop around for the best deal.

There was also a problem with *bounty jumpers*. Some ne'er-do-wells would join a unit that offered upfront cash bounties, then they never showed up or deserted to another part of the country and pulled the scam again, many times over.

These complications, along with the lack of central record keeping, make finding bounty records difficult and sometimes impossible.

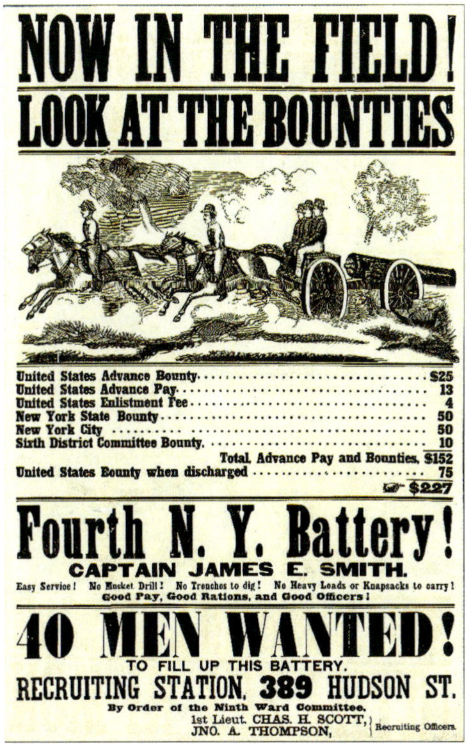

Bounty posters this page from Archive.org
https://archive.org/details/CivilWarPosters

Federal Bounty

Federal Bounty-Land Warrants were used in the Revolutionary War, the War of 1812, the Mexican War, and Indian Wars to entice or reward soldiers. These were typically in the form of *warrants* which were promises of a land grant to veterans. These bounties or warrants (terms used interchangeably) were initially successful, but fulfilling the promise was frequently not satisfactory. The practice of land bounties ceased in 1855. Searchable indexes for these are available on Ancestry and FamilySearch.

The Federal government resurrected bounties for Civil War enlistments, but with cash instead of land. A $100 bounty was approved by Congress in July 1861 for a three-year commitment. The *Enrollment Act* in March 1863 authorized a $300 bounty for three-year enlistments and $400 for five years, paid in monthly installments with a soldier's regular pay. The same act authorized a draft if a congressional district did not meet their enlistment quota.

The Federal bounty enticements were popular among many young men who faced a draft if they did not volunteer. The $300 bounty was quite large for the time; its equivalent buying power in 2021 dollars is more than $7,000.

Research Tip: There is NO central set of Federal Civil War bounty records. Land warrants (bounties) were authorized for the Revolutionary War and the War of 1812 only.

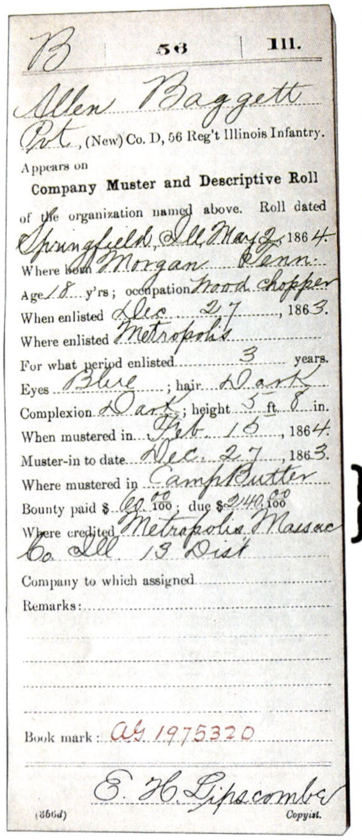

B | 56 | Ill.

Allen Baggett

Pvt., (New) Co. D, 56 Reg't Illinois Infantry.

Appears on

Company Muster and Descriptive Roll

of the organization named above. Roll dated

Springfield, Ill. May 2, 1864.

Where born Morgan Tenn.

Age 18 y'rs; occupation Wood chopper

When enlisted Dec. 27, 1863.

Where enlisted Metropolis

For what period enlisted 3 years.

Eyes Blue; hair Dark

Complexion Dark; height 5 ft. 8 in.

When mustered in Feb. 15, 1864.

Muster-in to date Dec. 27, 1863.

Where mustered in Camp Butler

Bounty paid $60 00/100; due $240 00/100

Where credited Metropolis Massac Co. Ill. 13 Dist

Company to which assigned

Remarks:

Book mark: AG 1975320

E. H. Lipscomb

(356d) Copyist.

Record Access

Evidence of a Federal bounty paid to a soldier is usually found in records like Muster Reports, discharge papers, or pay records. These types of reports are often found in a soldier's Compiled Military Service Record (CMSR).

This Muster Report of Allen Baggett shows a total of $60 of a $300 bounty was paid as of May 2, 1864, five months into his three-year enlistment. Subsequent reports show the paid amount was greater as payments were issued in increments over the term of the enlistment.

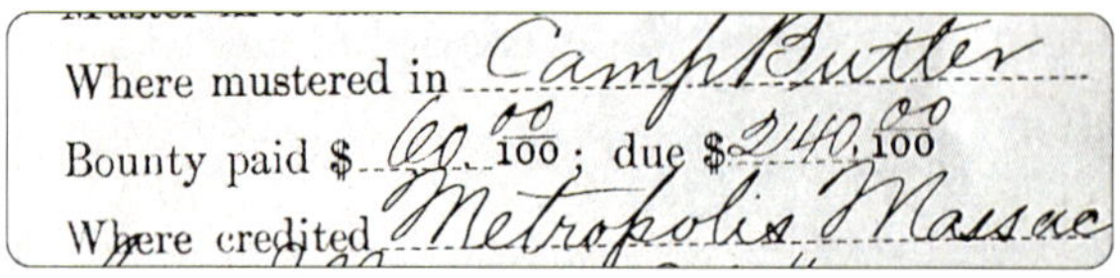

Local and State Bounties

State and local bounties were used to attract enlistments and satisfy federally mandated quotas. If a district fell short of filling an enlistment quota with volunteers, young men in the area could be drafted. Local bounties were generally smaller than Federal bounties. Yet, wealthy areas, not wanting their sons conscripted against their wishes, increased bounties to exorbitant rates to attract enlistments and fulfill the federal quotas.

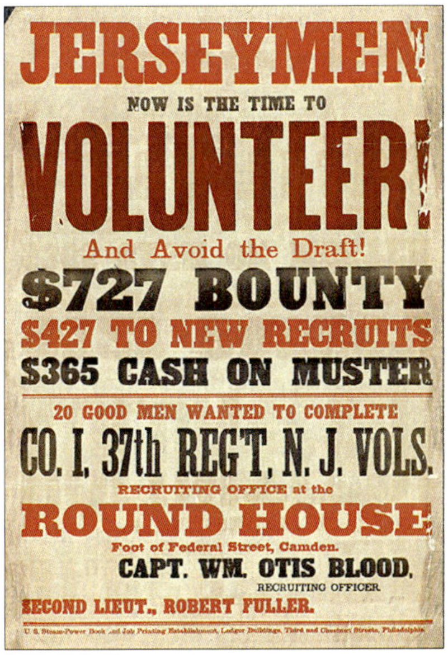

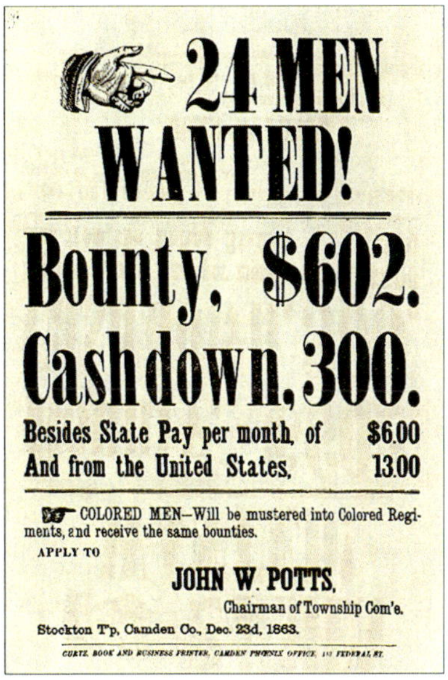

Bounty posters from Archive.org, https://archive.org/details/CivilWarPosters

The above posters offered locally funded bounties (which may have included a state bounty) in addition to the Federal payment, noting that by accepting these offers you would avoid the draft, which paid nothing. To further broaden the appeal, the poster on the right points out that "COLORED MEN" would be mustered in at the same rate.

A 1933 thesis titled *A History of the Bounty System Used During the Civil War* by Oren Ford covers the history of bounties, and how each state structured their bounties. However, it does not include names and details of individual recruits and soldiers.[4]

Another resource for bounty information is newspaper archives. Newspapers published many articles describing bounties during the Civil War along with bounty advertisements. The 1863 classified section in the December 15, 1864, *New York Daily Herald* listed over 40 ads placed by counties, bounty agents, and lawyers.

MILITARY AND NAVAL.

$600 TO VOLUNTEERS, ARMY OR NAVY. TAMMANY HALL. $75 cash to any person bringing a recruit. C. L. MORSE, Tammany Hall.

$600.—VOLUNTEERS AND SUBSTITUTES wanted.—Fifty good men to go as seamen or firemen in the navy, three substitutes in the army, this day. Capt. WM. RAINER, 73 Nassau street.

$600.—HOME SERVICE.—FEW GOOD MEN wanted for provost duty. Only honest, faithful men accepted, they having to guard others instead of being guarded. Apply at No. 2 Broadway, front office, second floor.

$600 BOUNTY.—VOLUNTEERS WILL DO WELL by calling on Captain FISHER, United States Army, 146 Chatham street, up stairs. $100 paid to any person bringing a recruit.

$700 FOR THREE YEAR VOLUNTEERS, $500 CASH. $450, all cash down, for one year substitutes. $400, all cash, for Invalid corps, and no humbug, at 97 Murray street.

$700 CASH BOUNTY TO VOLUNTEERS; $600 cash down to substitutes. Choice of service at 84 Broadway, room 8, up stairs.

$700.—TEN MEN WANTED THIS DAY—ARMY, navy or marine. $600 cash to substitutes. Choice of service. Apply early this day at the Metropolitan office, 81 Nassau street, room No. 4, first floor. FARRAR, JENKINS & CO.

$700 PAID EACH VOLUNTEER FOR THREE years; $500 cash down, besides government bounty, pay, &c. Apply early at CLARK & VOULES', 30 Fourth avenue, opposite Cooper Institute.

New York Daily Herald, New York, NY, 15 Dec 1864, p. 6

Record Access

Few records exist of local bounties, though some states organized their records. Chapter Eight includes bounty records that have been discovered by state.

Personal War Sketches

Some G.A.R. posts created a unique class of historical records. Probably as an outgrowth of regimental histories, numerous posts encouraged members to write their *Personal War Sketches*. Started as a grassroots program in the late 1880s, the practice was encouraged by the National G.A.R. At the 1891 National Encampment, the Adjutant General commented:

> *Many Posts of the country are engaged in procuring a biographic sketch of each of their respective Comrades. Such sketches will be a valuable acquisition, and the example which some Posts have set in the regard should become the rule throughout the Order.*[5]

Personal War Sketches book includes Amos C. Wertz's entry.

The practice started as a way to regain individual recognition in communities; sketches were presented at local meetings and schools. Some posts gathered them into folders or collections.

A Philadelphia map, atlas, and county history publisher, Louis H. Everts, produced *Personal War Sketches* books, with up to 100 blanks inside for veterans' personal information and Civil War experiences. Introduced in 1890, these were large books, approximately 12" x 17" that included the Grand Army of the Republic and patriotic themes embossed in gold leaf. Many entries were used as obituaries, others were entered after the veteran died. Still referenced today, Amos Wertz's sketch (next page) was used in his Find a Grave record.

These books were popular in Eastern posts. Fewer examples are found in the sparsely populated Plains and West, where many veterans migrated to claim homesteads.

Finding the Sketches

The truth is, these sketches did not cover a majority of G.A.R. members. The books are scarce, but some can be found in:

- Historical societies
- Genealogy libraries
- University archives
- Online archives

An Exceptional Sketch

Amos C. Wertz's Personal War Sketch is a fascinating and exceptional example. It appears written by another person, as many veterans could not write. His story included captivating detail.

Shortly after enlisting in August 1862, he was "captured by the enemy." He was released and discharged, then reenlisted in the U.S. Cavalry in January 1863, where he experienced more battles and a broken leg in August 1863. Returned to duty, he saw ten additional battles.

Wertz was discharged in 1865, but not before being "slightly wounded at Gettysburg," "bowled over by a spent shot" in October 1863, and badly burned from a "timber blown up." His gallantry was rewarded with promotions from private to corporal to sergeant to 1st Sergeant.[6]

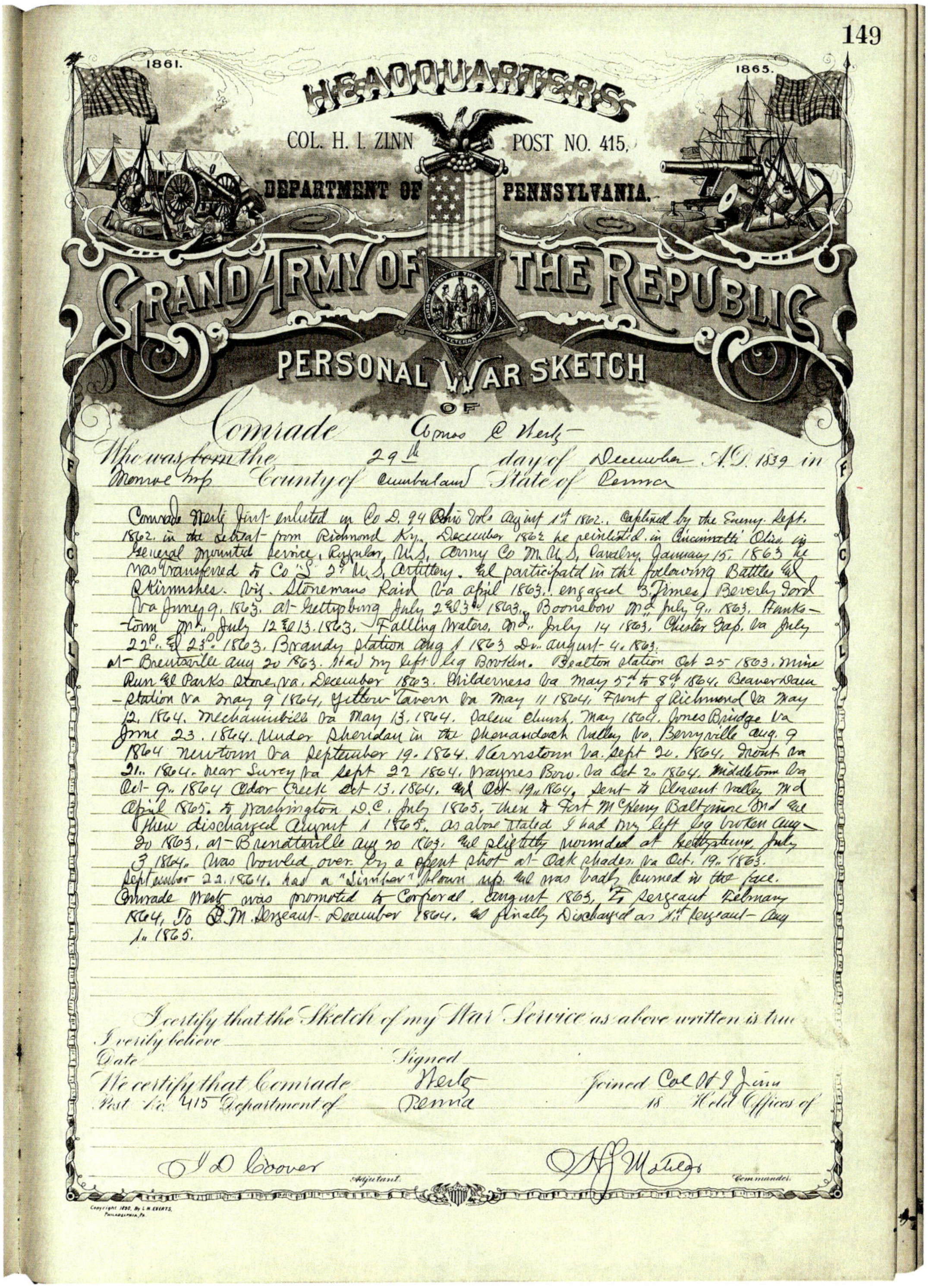
149

1861. 1865.

HEADQUARTERS

COL. H. I. ZINN POST NO. 415,

DEPARTMENT OF PENNSYLVANIA.

GRAND ARMY OF THE REPUBLIC

PERSONAL WAR SKETCH OF

Comrade Amos C Wertz

Who was born the 29th day of December A.D. 1839 in Monroe Twp County of Cumberland State of Penna

Comrade Wertz first enlisted in Co D. 94 Ohio Vols August 1st 1862. Captured by the Enemy Sept. 1862 in the retreat from Richmond Ky. December 1862 he reinlisted in Cincinnatti Ohio in General mounted service, Regular, U.S. Army Co M U.S. Cavalry. January 15, 1863 he was transferred to Co "L" 2d U.S. Artillery. and participated in the following Battles and Skirmishes. Viz. Stonemans Raid Va april 1863. engaged 3 Times. Beverly Ford Va June 9. 1863. at Gettysburg July 2d and 3d 1863. Boonsboro Md July 9. 1863. Hagerstown Md July 12 and 13. 1863. Falling Waters Md July 14 1863. Chester Gap Va July 22d and 23d 1863. Brandy Station Aug 1 1863 Do August 4. 1863.

at Brentsville aug 20 1863. Had my left leg Broken. Bealton Station Oct 25 1863. Mine Run and Parks Store, Va. December 1863. Wilderness Va May 5th to 8th 1864. Beaver Dam Station Va May 9 1864, Yellow Tavern Va May 11 1864, Front of Richmond Va May 12. 1864. Mechanicsville Va May 13. 1864. Salem Church. May 1864. Jones Bridge Va June 23. 1864. Under Sheridan in the Shenandoah Valley Va. Berryville aug. 9 1864 Newtown Va September 19. 1864. [illegible] Va. Sept 20. 1864. Front Va 21. 1864. near Surey Va Sept 22 1864. Waynes Boro. Va Oct 2. 1864. Middletown Va Oct 9. 1864 Cedar Creek Oct 13. 1864. and Oct 19. 1864, Sent to Pleasant Valley Md April 1865. to Washington D.C. July 1865. then to Fort McHenry Baltimore Md and then discharged August 1 1865. as above stated I had my left leg broken aug 20 1863, at Brentsville aug 20 1863. and slightly wounded at Gettysburg, July 3 1864. Was bowled over by a spent shot at Oak Shades Va Oct. 19. 1863. September 22. 1864. had a "Limber" blown up. and was badly burned in the face. Comrade Wertz was promoted to Corporal. August 1863, To Sergeant February 1864, To Q.M. Sergeant December 1864. and finally Discharged as 1st Sergeant Aug 1. 1865.

I certify that the Sketch of my War Service as above written is true I verily believe

Date Signed

We certify that Comrade Wertz joined Col H I Zinn Post No. 415 Department of Penna 18 Held Offices of

I D Coover Adjutant. [illegible] Commander.

Copyright 1890, By L. H. EVERTS, PHILADELPHIA, PA.

A Regimental History Reality

The Twentieth Connecticut, A Regimental History by John W. Storrs, of "Company Z," added the horrors of war to the facts and story. This excerpt (pages 70-72) finds its author telling the tragedy of the day and contributing mournful thoughts of the event.

The Twentieth Connecticut, A Regimental History.

. . . At 7 o'clock on the morning of the 18th, [June 1863] the march was resumed, with another excessively hot day, reaching Leesburg at 6 o'clock, p. m., where the men went into camp for the night. During the day, in addition to the intense heat, the men of the 12th had encountered tremendous hail storm and had waded Goose Creek with the water to their waists.

While the Corps lay at Leesburg, two men belonging to the 46th Penn. Vols., and one of the 13th New Jersey Vols., were shot to death as deserters. It was a hard fate, but as even the kind-hearted Lincoln, on appeal, refused to pardon, it must have been necessary. There were the usual terrible formalities. The ready excavated graves, the coffins at their brink, with the wretched prisoners and condemned seated thereon. There was the last convulsive grasp of the chaplain's hand; the long, last look at the sun ; the bandaging of the eyes, and the word "fire!" from the lips of the commanding officer and, perforated each with eight bullets, somebody's son, or brother, or father, tumbles over into his box, and the tragedy is ended. Yes, they were deserters, that is true. And for aught we know, good reader, "bounty jumpers." If so, they were, perhaps, rightly served.

But suppose that these unfortunates were only a trio of country lads that had been cajoled into enlistment by the recruiting sergeant who gave, as they generally did, the most glowing and picturesque accounts of army life; and suppose that, afterward, experiencing the reality as we have thus far seen the men of the 20th experiencing, a vision of the dear old home fireside on the far away northern hills had flashed upon their vision, and then in a moment of heart-sick despondency and weakness they had decided to make a break for home and liberty, stop fighting and suffering for others' happiness and look a little to their own? In this land of the free ought they to have been shot to death for the act? Military law answers "yes," and in war time that is the statute of all others that must be paramount, must be obeyed, because in no other way could armies be kept in the field. Without the occasional enforcement of the death penalty for offenses of this kind (and the number of cases were proportionately small) the men that after tattoo might decide not to appear at reveille would, perhaps, leave the faithful sometimes in a sad minority. . . [7]

8 State G.A.R. Record Sources

- State by State

Most original G.A.R. records are held in state and local agencies, universities, libraries, or private organizations such as historical societies. In addition, many states have compiled archives of soldier rosters, regimental histories, battle histories, bounty records, and other useful historical documents that pertain to the respective state. This chapter includes the more commonly known sources and some of their holdings.

State by state

G.A.R. and Civil War records residing in local or state archives each require a unique research path. In addition, many sources are not available online; they can only be researched in person. The entries denote whether they are online or onsite records. This list encompasses the major sources typically found; however, many smaller organizations such as local historical societies or libraries may have relevant collections also.

This appendix focuses on G.A.R. records. Search in the state or states where he was when he was in the G.A.R., not the state of his regiment (unless they are the same). Some of these sources may hold other Civil War records, in which case, relevant records to the veteran may be in his regiment's state.

Note: The Sons of Union Veterans of the Civil War website includes a *GAR by State* button. After clicking on a state, select *GAR Records Catalog*, which displays a table with known records by post. This does not contain actual records, but provides a detailed list where specific records are held. If you look at one source for G.A.R. records, make it this one.

http://suvcw.org/garrecords

Alabama

Alabama Department of Archives & History – 624 Washington Ave., Montgomery, AL
Alabama Civil War Service Database – Compiled from a multitude of sources at the Department of Archives and History, including muster rolls, newspapers, governors' records, veterans' censuses, and state pension records.
https://archives.alabama.gov/civilwar/

Alabama Department of Archives and History - Military Records
https://archives.alabama.gov/Research/military.html

Ancestry.com – Alabama, U.S., Confederate Pension and Service Records, 1862-1947
https://www.ancestry.com/search/collections/1593/

FamilySearch.org – Alabama, Confederate Pension Applications, ca. 1880-1930's
https://www.familysearch.org/search/collection/3029256

Alaska

Archives West – University of Washington Libraries, Special Collections, Seattle, WA
Grand Army of the Republic, Department of Washington and Alaska records.
http://archiveswest.orbiscascade.org/ark:/80444/xv98461

Arizona

Arizona Historical Society – 949 East Second St., Tucson, AZ
Grand Army of the Republic Finding Aid. Four page description of Arizona Historical Society holdings. Physical onsite search only.
https://arizonahistoricalsociety.org/

SUVCW Department of the Southwest
Sons of Union Veterans of the Civil War (SUVCW) - Arizona & New Mexico. Some excellent rosters listed under the Projects tab.
https://suvcw-southwest.org/

Arkansas

CALS Encyclopedia of Arkansas – Central Arkansas Library System(CALS) project
A gateway to many articles, records, and images of the G.A.R. and Civil War.
https://encyclopediaofarkansas.net/entries/grand-army-of-the-republic-2802/

Arkansas Digital Archives - A part of the Arkansas State Archives
https://www.arkansasheritage.com/arkansasstatearchives/home
An excellent 62-page Civil War resource guide (download).
https://digitalheritage.arkansas.gov/resource-guides/7/

California

California State Library - 900 N St. Sacramento, CA
No online records, but excellent collection of documents in library for onsite review. Search for 'Grand Army of the Republic' on the following page.
https://www.library.ca.gov/california-history/collections/

San Diego Historical Society - 1649 El Prado, Balboa Park, San Diego, CA
http://www.sandiegohistory.org/

University of CA, Riverside - 3401 Watkins Dr., Riverside, CA
Special Collections and Archives has original records for many California Posts.
http://library.ucr.edu/view/libraries/rivera

Bob Hope Patriotic Hall - 1816 S. Figueroa St., Los Angeles, CA
https://mva.lacounty.gov/bob-hope-patriotic-hall/

University of CA, Berkeley - Bancroft Library, University of CA, Berkeley, CA
Many original post records.
http://bancroft.berkeley.edu/

Colorado

Denver Public Library - Western History and Genealogy Department, 10 W. Fourteenth Ave Pkwy, Denver, CO
Largest Colorado collection of original post records.
www.denverlibrary.org/

Fort Lewis College, Center for Southwest Studies, Durango, CO

Original post records for several posts.

http://swcenter.fortlewis.edu/index.shtml

Connecticut

Connecticut State Library – 231 Capitol Avenue, Hartford, CT

RG 113, Grand Army of the Republic, Department of Connecticut, Inventory of Records. Includes general orders and circulars, correspondence, financial records, annual reports, post lists, membership lists, music, records relating to court martials of G.A.R. members, and scrapbooks. In addition, the collection includes records created by various G.A.R. posts throughout Connecticut. Includes Connecticut W.R.C. archives. Physical records only, not online.

https://ctstatelibrary.org/RG113.html#idp13600

Delaware

Delaware Public Archives - 121 Martin Luther King Jr. Blvd North, Dover, DE

Collection of original records from a small number of posts.

http://archives.delaware.gov/

Florida

No known independent collections of Grand Army records in Florida. Search Fold3, Ancestry.com, and FamilySearch.org. In 1890 Florida had only 414 G.A.R. members in 17 posts.

Georgia

No known independent collections of Grand Army records in Georgia. Search Fold3, Ancestry, and FamilySearch. In 1890 Georgia had only 432 G.A.R. members in 9 posts.

Hawaii

Bob Hope Patriotic Hall - 1816 S. Figueroa St., Los Angeles, CA

One descriptive book for Post 45 in Hawaii is available here.

https://mva.lacounty.gov/bob-hope-patriotic-hall/

Idaho

Idaho State Archives - 2205 Old Penitentiary Rd, Boise, ID

Large collection of department and post records. Onsite only.

http://history.idaho.gov/

Illinois

Illinois Historic Preservation Agency at the Abraham Lincoln Presidential Library - 112 North Sixth St, Springfield, IL

A large collection of department and post records (nearly 800) are available for research. Onsite only.

www.illinoishistory.gov/

Chicago History Museum - 1601 N. Clark St., Chicago, IL

Collection of many original post records. Onsite research only.

www.chicagohistory.org/

A 1914 roster of Illinois GAR members, 160 pages.

https://www.grantcamp.org/uploads/8/2/4/6/82468692/illinois_gar_roster.pdf

Illinois in the Civil War

A project to put Illinois Civil War rosters and history on the internet.

https://civilwar.illinoisgenweb.org/index.html

Indiana

Indiana Archives and Records Administration - Digital Archives

Search through the Indiana archives, choosing the military category. Volunteers entered more than 213,000 names of Indiana veterans into this searchable database. The information on service was culled from the muster rolls of Indiana's infantry, cavalry, and battery units to assist soldiers and sailors in obtaining their pensions.

https://secure.in.gov/apps/iara/search/

Allen County Public Library - Ft. Wayne, IN

Collection includes original records of several posts records. See the SUVCW.org site (listed at beginning of chapter) for detailed listings.

www.acpl.lib.in.us/

Indiana Historical Society – 450 West Ohio Street, Indianapolis, IN

https://indianahistory.org/

SUVCW, Department of Indiana

SUVCWIN.org – Sons of Union Veterans of the Civil War, Indiana Department

http://suvcwin.org/?page_id=573

Iowa

Family Search – Index cards transcribed from G.A.R. records

Iowa, Grand Army of the Republic Membership Records, 1861-1949.

https://www.familysearch.org/search/collection/2351982

Ancestry – Index cards transcribed from G.A.R. records

https://www.ancestry.com/search/collections/62043/

Iowa Civil War Monuments – Online only

Listing of Iowa monuments, G.A.R. halls, and cannons. Photos and descriptions.

https://www.iowacivilwarmonuments.com/

State Historical Society of Iowa - Des Moines Research Center, State Historical Building

Adjutant General holdings index. Physical records only.

https://iowaculture.gov/history/research/collections/state-archives/rg-101-adjutant-general

Kansas

Kansas State Historical Society – 6425 SW 6th Avenue, Topeka KS

www.kshs.org

Kansas Adjutant General's Report, 1861-1865.

https://www.kshs.org/p/kansas-adjutant-general-s-report-1861-1865/11175

Family Search – www.familysearch.org

Journals of Annual Kansas Encampments – 1913-1919.

https://www.familysearch.org/search/catalog/2568598

Ancestry – www.ancestry.com

Members and post roster, 1894, Grand Army of the Republic, Dept of Kansas.

https://www.ancestry.com/search/collections/1700/

Kentucky

Only scattered records throughout the state. See SUVCW.com described at beginning chapter for access to details,

Louisiana

Tulane University, Howard-Tilton Memorial Library - 7001 Ferret St., New Orleans, LA

A collection of department records, onsite only.

http://library.tulane.edu/

Louisiana State University, LSU Libraries - Baton Rouge, LA

A collection of post records, onsite only.

https://www.lib.lsu.edu/

Maine

Maine State Library - 230 State St, Augusta, ME

Variety of department and post records.

https://www.maine.gov/msl/

Maine Historical Society - 489 Congress St., Portland, ME

Collection of post records.

https://www.mainehistory.org/

SUVCW.org - See SUVCW.org/garrecords for many small collections.

Maryland

Maryland State Archives -350 Rowe Boulevard, Annapolis, MD

Civil War Research in Maryland. Great list and index of records, onsite only.

http://guide.msa.maryland.gov/pages/viewer.aspx?page=civilwar

Maryland Historical Society - 201 West Monument St., Baltimore, MD

Various department and post collections.

http://www.mdhs.org

Massachusetts

New England Historical Genealogical Society (NEHGS) - 99–101 Newbury Street, Boston, MA

Both a physical repository for G.A.R. and Civil war records and an online database with Civil War records and extensive genealogy and historical collections.

Note: American Ancestors is the website of the NEHGS.

https://www.americanancestors.org/

Massachusetts Historical Society, 1154 Boylston Street, Boston, MA

Search the Library Catalog for Grand Army to reveal thousands of records, ledgers, journals, photographs, and other documents. Mostly onsite, some online.

http://www.masshist.org/

Historic New England - 141 Cambridge Street, Boston, MA

Many photographs, books, documents, & papers. Search collections for Grand Army.

https://www.historicnewengland.org/

Grand Army of the Republic Hall and Museum in Lynn, MA

A museum set in period style with a great number of artifacts.

https://garlynn.org

Hingham's Grand Army of the Republic Memorial Hall, Hingham, MA

Hingham's G.A.R. Memorial Hall is used by veterans for meetings and special events for Memorial and Veterans Days. The G.A.R. Hall also houses historic and wartime memorabilia.

https://www.facebook.com/Hinghams-Grand-Army-of-the-Republic-Memorial-Hall-1180924048595573

Michigan

Michigan Library and Historical Center - 702 W. Kalamazoo St., Lansing, MI

Extensive collection of department and post records.

http://www.michigan.gov/michiganhistory

Minnesota

Minnesota Historical Society - 345 W. Kellogg Blvd, St. Paul, MN

Military Service Records Research: Civil War Records. Includes Finding Aids, which link to documents identifying records in the society. Records include Adjutant General Reports, Military Service Records, Regiment Rosters, and others.

https://libguides.mnhs.org/milservice/cw

Mississippi

No known independent collections of G.A.R. records in Mississippi. Search Fold3, Ancestry.com,and FamilySearch.org. In 1890, there were no Mississippi G.A.R. posts.

Missouri

Missouri State Archives - 600 West Main St., Jefferson City, MO

Guide to Civil War Resources at the Missouri State Archives.

https://www.sos.mo.gov/archives/resources/civilwar

U.S. Grant Camp

A Sons of Veterans of the Civil War camp.

https://www.grantcamp.org/

University of Missouri Libraries

U.S. Civil War in Government Documents

https://libraryguides.missouri.edu/civilwargov

Montana

Montana Historical Society Archives - 225 North Roberts, Helena, MT

Extensive collection of department and post records in Archival Collections.

https://mhs.mt.gov/research/collections/archive

Nebraska

Nebraska State Historical Society -1500 R Street, Lincoln, NE.

A fantastic collection of G.A.R. and Civil War records, most only in physical form, requiring onsite research. See the about page for detailed contact information. Search for 'Grand Army' or 'Civil War' to retrieve lists of relevant material.

https://history.nebraska.gov/

Nebraska Civil War Veterans Database

A searchable database, if name is found it returns Unit Served, GAR Post No., GAR Post City, GAR Post County.

https://history.nebraska.gov/collections/nebraska-civil-war-veterans-database

FamilySearch

Search the Wiki for 'Nebraska in the Civil War', the results will include "Nebraska, Grand Army of the Republic, Burial Records, 1861-1948." Online.

Nevada

No known independent collections of Grand Army records in Nevada. Search Fold3, Ancestry, and FamilySearch. In 1890 Nevada had no G.A.R. posts.

New Hampshire

New Hampshire Historical Society - 30 Park St, Concord, NH

Extensive collection of department and post records.

https://www.nhhistory.org/

SUVCW.org - See SUVCW.org/garrecords for many small collections.

New Jersey

New Jersey State Archives – 225 West State Street - 2nd Floor, Trenton, NJ

Enlistment, Muster, Bounty, Pension, Battles, and many other records, onsite only.

https://www.state.nj.us/state/archives/catsedefen4.html

SUVCW.org - See SUVCW.org/garrecords for many excellent small collections.

New Mexico

New Mexico Archives & Records Center -1205 Camino Carlos Rey, Santa Fe, NM

Excellent collection of department and post records.

https://www.srca.nm.gov/

New York

New York State Archives - Records of Military Service, Online only.

You can search a database of New York soldiers who served in the Civil War and learn more about other Civil War records held by the New York State Archives.

http://www.archives.nysed.gov/research/res_topics_gen_guide_mil.shtml

Ancestry – Online Only - Scanned and indexed G.A.R. records

An excellent extensive collection of New York G.A.R. records. Indexed for Name search, and scans listed by Post Numbers. Scans of Post Descriptive Books, circulars, notices of National Encampments, war songs, order books, accounting materials, minutes, death and burial registers, and post rules.

https://www.ancestry.com/search/collections/5358/

North Carolina

No known independent collections of Grand Army records in North Carolina. Search Fold3, Ancestry, and FamilySearch. In 1890, there were no N.C. G.A.R. posts.

Ohio

Ohio History Center - 800 E. 17th Ave, Columbus, OH

Collection of a department few post records

https://www.ohiohistory.org/learn/archives-library

SUVCW.org - See SUVCW.org/garrecords for many small collections.

Oklahoma

Oklahoma Historical Society - 800 Nazih Zuhdi Dr., Oklahoma City, OK

Research Center - Military Records. Includes Confederate rosters and pensions as well as Union records.

https://www.okhistory.org/research/military

Oregon

Oregon Historical Society - 1200 SW Park Ave, Portland, OR

A very good collection of department and post records.

https://www.ohs.org/

FamilySearch.org - Online

The same records in the Oregon Historical Society are available on microfilm and digitized at FamilySearch.

https://familysearch.org/search/collection/2239231

Oregon Secretary of State - 900 Court St NE, Salem, OR

The State Archives offers Oregon Military Department records for research. These are for in-person research, not online.

https://sos.oregon.gov/archives/Pages/records/aids-military.aspx

Pennsylvania

Pennsylvania State Archives – 3rd & Forster, Harrisburg, PA.

A good physical collection of GAR charters, Descriptive Books, and miscellaneous post records; access onsite only. List of holdings:

http://www.phmc.state.pa.us/bah/dam/mg/ys/m60ys.htm/.

https://www.phmc.pa.gov/

University of Pittsburgh

Guide to the Grand Army of the Republic, Dept. of Pennsylvania, Miscellaneous Posts Records, 1861-1940.

https://digital.library.pitt.edu/

Chester County Pennsylvania – Archives and Record Services

The archives include a log of *Civil War Bounty Records, 1862-1865,* including individual names, some details of their bounty, and where to find the original record.

https://www.chesco.org/1703/Civil-War-Bounty-Records-1862-1865

FamilySearch – Online Only - Scanned G.A.R. records

Pennsylvania, Grand Army of the Republic Membership Records, 1861-1949. These records, and the Ancestry records, are G.A.R. record scans. The Name search works well, but browsing the scanned records is difficult as groups of records (posts) were not separated. They are simply listed by Box #, not by town or post number.

https://www.familysearch.org/search/collection/2207723 (Name Search)

https://www.familysearch.org/search/image/uri=https://www.familysearch.org/recapi/sord/collection/2207723/waypoints (Browse record books)

Ancestry – Online Only - Scanned G.A.R. records

See previous note with Family Search.

https://www.ancestry.com/search/collections/61877/

Rhode Island

Rhode Island Historical Society - 121 Hope St., Providence, RI

Extensive collection of department and post records.

www.rihs.org/

South Carolina

No known collections of Grand Army records in S. Carolina. Search Fold3, Ancestry, and FamilySearch. In 1890 South Carolina had no G.A.R. posts.

South Dakota

South Dakota Historical Society - 900 Governors Drive, Pierre, SD

Holds a large collection of department and state original records, journals, et al.

http://history.sd.gov/

FamilySearch

Excellent online collection of images from microfilm of the original records held by S.D. Historical Society. Includes membership rosters, attendance registration books of various encampments (some include Women's Relief Corps.), post descriptive books, member deaths, adjutant reports, muster rolls, and others.

https://www.familysearch.org/search/collection/2239227

Tennessee

Knox County Public Library - 500 West Church Avenue, Knoxville, TN

Limited collection of department and post records.

https://www.knoxlib.org/

Texas

No known collections of Grand Army records in Texas. Search Fold3, Ancestry and FamilySearch.

Utah

Utah State Historical Society - 300 S. Rio Grande St., Salt Lake City, UT

Extensive collection of original department and post records, many available online from FamilySearch.

http://history.utah.gov/

FamilySearch.org

This online collection consists of registers, lists and descriptive books of local posts from the Utah State Historical Society.

https://www.familysearch.org/search/collection/2203361

Vermont

Vermont Historical Society – Barre, VT

Original G.A.R. journals and records. W.R.C. journals. No online record access, onsite access only.

http://vermonthistory.org/research/leahy-library.

University of Vermont Libraries – Bailey/Howe Library, Burlington, VT

Original G.A.R. journals and reports, W.R.C. journals. In Special Collections.

http://library.uvm.edu/

Sons of Union Veterans of the Civil War (SUVCW)

Lists of posts by state with some historical information, and sources of information.

http://suvcw.org/garrecords/

Vermont in the Civil War – A virtual organization.

A crowd sourced website with collections of Civil War records and photographs.

https://vermontcivilwar.org/

Virginia

Library of Virginia - 800 East Broad St., Richmond, VA

Nice collection of G.A.R. records. Search the LVA Catalog for 'Grand Army of the Republic'.

https://www.lva.virginia.gov/

Washington

Washington State Library - Point Plaza East 6880 Capitol Blvd SE, Tumwater, WA

Collection of department and post records.

www.sos.wa.gov/library

Washington State Archives - 1129 Washington St., Olympia, WA

Collection of post records.

http://www.sos.wa.gov/archives/archives_state.aspx

West Virginia

West Virginia University - 1549 University Ave., Morgantown, WV

West Virginia & Regional History Center

https://wvrhc.lib.wvu.edu/research/subject-guides/military

Wisconsin

Wisconsin Veterans Museum - 30 W. Mifflin St., Madison, WI

Collection of a few post records.

https://wisvetsmuseum.com/

Wisconsin Historical Society - 816 #rd St. South, Wisconsin Rapids, WI

Collection of a few post records.

https://www.wisconsinhistory.org/

University of Wisconsin - Various Locations

Post records are scattered among university campuses around the state. See the SUVCW.org/garrecords webpage for a detailed listing.

https://suvcw.org/garrecords/

Wyoming

Denver Public Library - 10 W. Fourteenth Ave. Pkwy, Denver, CO

Most Wyoming records are in the Denver Public Library.

https://www.denverlibrary.org/

University of Wyoming, American Heritage Center - 100 E. University Ave, Laramie, WY

Collections of only a couple posts.

https://www.uwyo.edu/ahc/

Appendix A – Regiment Types

Most veteran records are distinguished by the state where they enlisted and their assigned regiment. G.A.R. records nearly always included a member's regiment and company, or ship name in muster rolls, descriptive books, adjutant reports, rosters, and encampment journals. Knowing the regiment and company will help identify a specific veteran among others with a similar name. Also, it is helpful for the researcher to understand the role of a veteran's regiment as it may add historical context about their service. If a veteran was in the Navy, his ship is used to designate unit affiliation.

Below are examples of the most common regiment designations, though unique abbreviations are frequently found on documents and gravestones. Space constraints, or the writer or engraver's inclination, led to many non-standard variations.

Regiment Type	Abbreviations	Example/Note
Infantry		
Infantry	Infantry, INF	7th Ohio Infantry
Volunteer Infantry	Infantry, INF, V.I.	11th Michigan Volunteer Infantry
Mounted Infantry	Mt Inf., Mtd Inf.	(Infantry on horseback)
Mounted Rifles	M. R., Mtd Rifles	1st N.Y. Mounted Rifles, (Same as Mtd Inf)
Cavalry		
Cavalry	Cav.	11th PA Cav
Volunteer Cavalry	V.C., Vol Cav.	1st Vol Cav

Artillery

Light artillery (field) pieces were moved by teams of horses and were deployed on the battlefield in both offensive and defensive situations. Heavy artillery was used in fixed areas such as forts, towns, or harbors, or in longer term siege attacks against forts and towns.

Regiment Type	Abbreviations	Example/Note
Artillery	Artillery, Art, Arty	7th NY Artillery
Heavy Artillery	H.A., H., Hv. Art.	5th NY H. Art
Light Artillery	Lt. Art., L.A.	1st Ohio Lt. Art
Battery	Battery, Batt.	Battery B (Smaller unit within a regiment)
Miscellaneous		
Vol. Reserve Corps	V. R.C., R.C.	15th V. R.C. (Partially disabled or infirm)
Independent	Ind. Battery B.	Typically a company or battery size unit
Militia	Mil., Militia	26th PA Mil Inf
Sharpshooters	Sharpshooters	Usually company units named after commander, e.g. Beatty's Sharpshooters.
Colored	U.S.C.T., US Col	18th U.S.C.T. Inf., 59th US Col., 122 U.S.C.I.

Examples are from muster rolls or gravestones. Gravestones on the next page are captioned with the full regiment name represented by the abbreviation on the stone. Note: Spanish American War gravestones look like Civil War versions as these were allowed to use the engraved *shield*, along with Revolutionary War, War of 1812, and Indian Wars gravestones. Subsequent war gravestones were not engraved with the shield.

2d Colored Troops Heavy Artillery

5th Tenn. Mounted Infantry

11th Indiana Cavalry

26th Pennsylvania Militia Infantry

2d Missouri Light Artillery

137th Illinois Infantry

12th Penn. Reserve Infantry

3d New York Light Artillery

17th U.S. Colored Infantry

2d Minnesota Infantry
Medal of Honor Recipient

U. S. Navy

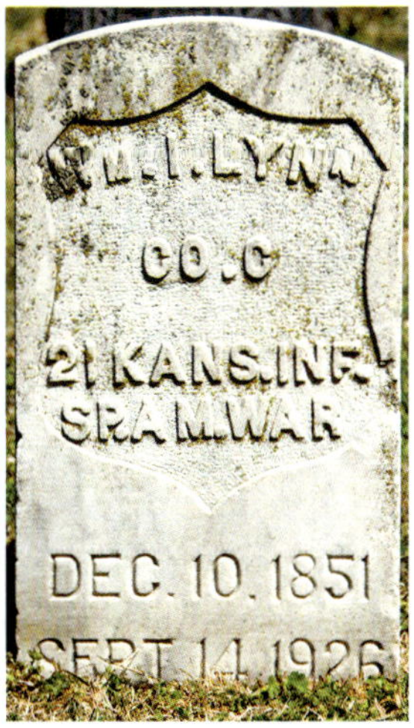

21st Kansas Infantry
Spanish American War

Appendix B – Pension File Numbering

Civil War pension files were numbered to help identify, store and locate them among the millions archived in several places. As the Pension Bureau was integrated into what was to become the Veterans Administration in the 1930s, some folders were archived to NARA in Washington D.C., others were sent to the St. Louis NARA, and active files were transferred to the Veterans Administration (renamed Veterans Affairs in 1988). In 1928, the numbering system was changed to indicate file status, which eventually led to an indication of the file's location.

This file number is on the pension folder tab, and may appear on a Pension Index Card. The Index Card was described in Chapter Six, though this appendix explains further the numbering system and may help reveal were the file is located.

Number Series

Before 1928, the pension certificate number was used as the file number. A prefix denoted the status or type of pension. Files closed before 1928 are stored using these prefixes:

- SO - Original filing series for Wounds or Illness pension, replaced if SC was granted.
- SC - Soldier's Certificate number issued when pension certificate granted.
- WO - Widow filing, replaced if WC was granted.
- WC - Widow's Certificate number when certificate granted.
- Prefixes also included MO and MC for minors, and R for civilians.

After 1928 the number system was changed for new and active pensions. It included a new five or six digit numeric file number along with one of the following prefixes:

- C - Active veteran pension.
- XC - Changed from "C" when veteran died, may include surviving widow/minor pension.

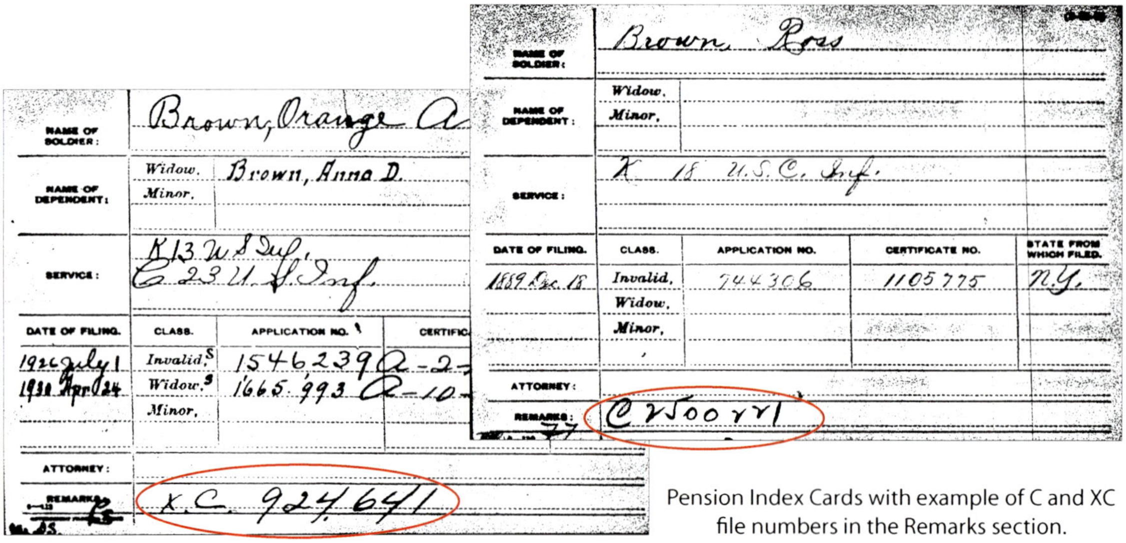
NAME OF SOLDIER: Brown, Ross
NAME OF DEPENDENT: Widow, Minor,
SERVICE: K 18 U.S.C. Inf.
DATE OF FILING: 1889 Dec 18 | CLASS: Invalid | APPLICATION NO.: 744306 | CERTIFICATE NO.: 1105775 | STATE FROM WHICH FILED: N.Y.
ATTORNEY:
REMARKS: C 2500771

NAME OF SOLDIER: Brown, Orange A.
NAME OF DEPENDENT: Widow, Brown, Anna D.
SERVICE: K 13 U.S. Inf. / C 23 U.S. Inf.
DATE OF FILING: 1926 July 1 | CLASS: Invalid | APPLICATION NO.: 1546239 | A-2
DATE OF FILING: 1930 Apr 24 | CLASS: Widow | APPLICATION NO.: 1665993 | A-10
ATTORNEY:
REMARKS: X.C. 924641

Pension Index Cards with example of C and XC file numbers in the Remarks section.

No File Number Location

Many Index Cards do no list an explicit file number. Before 1928, the Certificate Number was used as the file number. A missing file number in Remarks, usually means the veteran died before 1928 and the file was moved to NARA in Washington D.C. However, if it included a living widow or minor, it was moved to the Veterans Administration as an active file. Upon death of last pensioner the file was then moved to NARA.

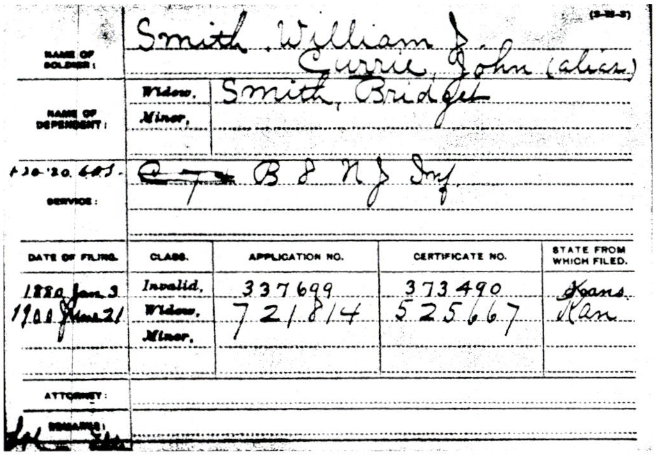
NAME OF SOLDIER: Smith, William J. / Currie, John (alias)
NAME OF DEPENDENT: Widow, Smith, Bridget / Minor,
SERVICE: C 7 B 8 N.J. Inf.

DATE OF FILING.	CLASS.	APPLICATION NO.	CERTIFICATE NO.	STATE FROM WHICH FILED.
1880 Jan 3	Invalid,	337699	373490	Kans.
1900 June 21	Widow,	721814	525667	Kan.
	Minor,			

ATTORNEY:
REMARKS:

William Smith died in 1900. Bridget Smith pension received his pension until her death in 1932.

C and XC Numbered File Locations

Most pension files are located in Washington D.C. NARA or in St. Louis NARA. The location is important to know if you are planning on visiting the holding-archive to view the file. The following notation in the Remarks section should identify the location of the majority of files.

- No file number listed – Washington, D.C. Archive
- C prefixed number – Washington, D.C. Archive
- XC prefixed number – St. Louis NARA Archive

The following is from a blog post by National Archives on history.gov (See reference below):

> *The National Archives at St. Louis (RL-SL) has custody of pension records ranging from XC 2 to XC 3990713. It is important to remember that because of how the cases close, several cases were still open at the time of transfer, causing gaps in our record series. These files are arranged by their XC number and files closed as of 1955 have been accessioned into the National Archives . . .*

Despite the designations and procedures, many pension records seem missing; stories about files not found by NARA are frequent. Do not let this stop your search.

> **Research Tip**: Include the file number on a record request. Use the Certificate No. if no file number is listed in Remarks.

Reference

The article *Anatomy of a Union Civil War Pension File,* by Claire Prechtel-Kluskens, National Archives, is an excellent overview of pension history, file organization, and location information.
https://www.archives.gov/files/calendar/genealogy-fair/2010/handouts/anatomy-pension-file.pdf

Seeking explanation of Civil War Pension XC number and A-certificate by NARA. September 15, 2020
https://historyhub.history.gov/thread/8500

C and XC Pension Files for the Civil War by Matthew Schaefer, May 14, 2010.
https://narations.blogs.archives.gov/2010/05/14/c-and-xc-pension-files-for-the-civil-war/

Appendix C - Glossary

Many researchers are not familiar with military and Grand Army terminology. This glossary includes terms from both.

Adjutant	Assistant to the commander, typically the top operational officer in an organization. In charge of record keeping/reporting. Applies to military, G.A.R., and government, sometimes referred to as Adjutant General.
Allied Associations	Organizations with same goals of or offering assistance to G.A.R. but not comprised of veterans. These were formally recognized by the G.A.R.
Artillery	Commonly thought of as cannons, both portable (light) and fixed (heavy).
ASUVCW	Auxiliary to Sons of Union Veterans of the Civil War is a women's auxiliary to the Sons of Union Veterans of the Civil War (SUVCW).
Auxiliary	A unique designation given to the Woman's Relief Corps; as in "The Official Auxiliary to the G.A.R." Other organizations were "Allied."
Battery	A small unit (aproximately150 men. can vary) with cannons in an artillery regiment. Typically 4 - 6 batteries in a regiment.
Blue Book	Formally titled *Rules and Regulations*, was blue and the cover included *Blue Book.*
Bounty	Cash bonus (whole or installments) as enticement to join the service during the Civil War. Offered by U.S., states, and local governments.
Cavalry	Soldiers mounted on horses and capable of firing and fighting from horseback.
Chaplain	Post and headquarters level officers, the spiritual leader of events and ceremonies.
Charter	An ornate certificate granting a new post formal standing in the G.A.R. It was surrendered to the department when the post was disbanded.
Company	A small unit (about 150 men) in an infantry unit. Typically 4 - 6 companies in a regiment. Number of men can vary greatly.
Commander	Top officer in a G.A.R. post, department and national headquarters.
CMSR	Compiled Military Service Record. File of a soldier's military records, stored in the National Archives with some available online.
DUVCW	Daughters of Union Veterans of the Civil War was a ladies organization, lineage to a veteran was required.
Department	A state or territory headquarters of the G.A.R. All posts report to the department, it reports to national.

Encampment	Annual multi-day gathering of G.A.R. members. Discussed business, passed resolutions, elected officers, and enjoyed camaraderie. These were held at the National and department level each year.
Encamp. Journal	A published book of the encampment proceedings. Included some rosters.
General Orders	Published communications from the National Headquarters and departments which invoked procedures, rules, declarations, and sometimes event announcements.
Infantry	Fighting foot soldiers. May have traveled between battles on wagons, trains, or boats.
LGAR	Ladies of the G.A.R., an allied association to the G.A.R.
Muster	A collection or gathering of troops. A G.A.R. Muster Roll is a list of all troops present. A Civil War Muster Roll included those missing, with noted reason such as AWOL, deserted, captured.
Pensions	Monthly payments to veterans based on injury and/or service. Also available to widows and minor children of deceased soldiers.
Post	A local (and smallest) unit of the G.A.R., organized at the town, city, or township level. Only ten veterans were needed to apply for a post charter.
Quartermaster	Officer in charge of funds, forms, etc. Supervised any commercial activities such as purchasing.
Regiment	A fighting force of approximately 600 - 1,000, though actual number can vary widely. Most Civil War regiments were raised by states.
Ritual	Specific and structured ceremonies were outlined in the Ritual Manual, for meetings, opening ceremonies, officer reports, ballot on recruits, muster-in (initiation), and others.
Surgeon	Post officer, not a medical person. He collected information about and reported conditions of members' Civil War wounds to the department. This information was important to expanding pension programs.
SUVCW	Sons of Veterans of the Civil War, a men's allied organization to the G.A.R. Member were direct descendants of Union veterans.
Troop	A small unit (about 150 men) in a cavalry regiment. Typically 4 - 6 troops in a regiment. Equivalent organization level to company (Infantry) or battery (Artillery). Number of men varied.
W.R.C.	Woman's Relief Corps, (Women's) Auxiliary to the G.A.R.

Appendix D – Research Checklist

The following steps should be accomplished in order to assist your G.A.R. and Union veteran research. Some items yield quicker results than others. This order will optimize your efforts.

Grand Army of the Republic Members

- ☐ Determine state enlisted, regiment, company, and rank of your veteran.
  - ☐ National Parks Service database lookup: www.nps.gov/civilwar/soldiers-and-sailors-database.htm.
  - ☐ Ancestry Civil War lookup: www.ancestry.com/cs/civilwarrecords.
- ☐ Determine state where he joined the Grand Army. G.A.R. records are organized by the state of the post they joined, NOT the state where they enlisted in the army (if different).
  - ☐ Where did they live after the war? City, town, county; especially 1880 - 1910.
  - ☐ Check www.suvcw.org/garrecords for list of posts by state.
- ☐ See Chapter 8, State Resources for online and local post G.A.R. records.
- ☐ Search newspapers for articles of post activity in their area, especially 1880 - 1920.

Union Service and Pension Records

- ☐ Determine state enlisted, regiment, company, and rank as listed above.
- ☐ Look for Compiled Military Service Record outlined in Chapter 6, Federal Records.
- ☐ Look for Pension Card and Records as outlined in Chapter 6.
- ☐ Look for Civil War Headstone Request as outlined in Chapter 6.

State Bounty, Adjutant Report, Regimental History

- ☐ Determine state enlisted, regiment, company and rank as listed above.
- ☐ Look for Regimental Histories, Adjutant General Reports, Bounty Records, and Personal War Sketches as outlined in Chapter 7, State and Local Records.
- ☐ Regimental History
  - ☐ Regimental History - Search Internet Archive (archive.org) using state, regiment, and the term 'regimental history'.
  - ☐ Search sources listed in the "Regimental History Access" section in Chapter 7.
  - ☐ Conduct general internet search for state, regiment, and 'regimental history'.
- ☐ Adjutant General Reports - Search libraries, state archives, historical societies for state and 'Adjutant General Report'.
- ☐ Bounty Records - Conduct an internet search for each state and 'Civil War bounty'.
  - ☐ Look for bounty payment notes on muster reports and other documents in a veteran's CMSR.

Notes & Recommended Reading

Chapter 1 - History

1.Tim Pinnick, "African American Veterans in the Grand Army of the Republic," *NGS Magazine* 35 No. 3 (July-September 2000), 28-32.

2. Beath, Robert B., *History of the Grand Army of the Republic* (New York, Bryan, Tailor & Co. 1888), 33.

3. Carnahan, J. Worth, *Manual of the Civil War and Key to the Grand Army of the Republic and Kindred Societies,* (Washington D.C., U.S. Army & Navy Historical Association, 1899), 18.

4. Beath, Robert B., *History of the Grand Army of the Republic* (New York, Bryan, Tailor & Co. 1888), 34-35.

5. *The Grand Army Blue Book Containing The Rules and Regulations of the Grand Army of the Republic* (Philadelphia, Burk & McFetridge Printers, 1884), 4.

6. ibid. p. 12.

7. *The Eagle*, Bryan, Texas, September 9, 1898, Page 2. Accessed Newspaper.com July 4, 2021

8. T*he Grand Army Blue-Book Containing The Rules and Regulations of the Grand Army of the Republic* (Philadelphia, Burk & McFetridge Printers, 1884), 218.

9. Beath, Robert B., *History of the Grand Army of the Republic* (New York, Bryan, Tailor & Co. 1888), vii.

10. Dearing, Mary R., *Veterans in Politic*s (Baton Rouge, LSU Press 1952), 1.

11. *Proceedings of the of the National Encampment of the Grand Army of the Republic*, (Washington, 1870), 11.

12. *Proceedings of the Eighth Annual Meeting of the National Encampment*, (Boston, Headquarters Grand Army of the Republic 1874), 21.

13. McConnell, Stuart, *Glorious Contentment: The Grand Army of the Republic*, 1865-1900, (Chapel Hill, The University of North Carolina Press), 32.

14. Gladwell, Malcolm. "The Stickiness Factor." The Tipping Point: How Little Things Can Make a Big Difference. (New York: Little, Brown and Co., 2000), 89-132.

15. *Ritual of the Grand Army of the Republic* (Zanesville, Ohio, Headquarters Grand Army of the Republic 1907), 15.

Chapter 2 - Research

1. National Archives, https://www.archives.gov/research/military/civil-war/resources, last accessed June 26, 2021.

2. *Journal of the 22nd Annual Session of the National Encampment Grand Army of the Republic* (Minneapolis, Harrison & Smith, Printers), 32.

3. *Journal of the 22nd National Encampment*, 1888, Adjutant Generals Report, 33.

4. Beath, Robert B., *History of the Grand Army of the Republic* (New York, Bryan, Tailor & Co. 1888), 44.
5. *Journal of the 22nd National Encampment,* 1888, Adjutant Generals Report, 60.
6. 1875 *9th National Encampment Journal*, 13.
7. *Journal of the Fortieth National Encampment*, 1906.

Chapter 3 - Affiliate Organizations

1. Red Book - *A Manual Containing the Rules and Regulations of the Woman's Relief Corps* (Boston, E.B. Stillings & Co., 1900), vi.
2. *Journal of the Fifteenth National Convention of the Woman's Relief Corps 1897*, Buffalo, NY, (Boston, E.B. Stillings & Co, 1897), 404.
3. 1*7th Annual Session of the National Encampment Grand Army of the Republic at Denver,* CO 1883 ((Omaha, NEB, Republican Book and Job Printing House, 1883), 793.
4. *Journal of the Twenty-Third National Convention of the Woman's Relief Corps 1905*, Denver, CO, (Boston, Griffith-Stillings Press, 1905), 361.
5. *Journal of the Twenty-Third National Convention of the Woman's Relief Corps* (Washington D.C., The National Tribune Company, 1917), 467.
6. *Regulations for the Government of the Woman's Relief Corp*, 1894.
7. *Sons of Veterans Blue Book - Digest of Decisions on the Constitution, Rules and Regulations of the Sons of Veterans USA*, (Indianapolis, Wm. B. Burford,1890), 8.
8. Ibid, p. 10

Chapter 4 - African American G.A.R.

1. *Preserving the Legacy of the United States Colored Troops*, National Archives, https://www.archives.gov/education/lessons/blacks-civil-war/article.html.
2. *Black Soldiers in the U.S. Military During the Civil War*, National Archives, https://www.archives.gov/education/lessons/blacks-civil-war.
3. Gannon, Barbara A., *The Won Cause: Black and White Comradeship in the Grand Army of the Republic* (Chapel Hill, The University of North Carolina Press 2011), 203.

Chapter 5 - Badges, Monuments & Cannons

1. Johnson, Kenneth R and Floyd, Jeffrey B., *Membership and National Encampment Badges of the Grand Army of the Republic* (Orders and Medals Society of America, 1997).
2. The Monumental Bronze Co. 1882 catalog, p. 18, Bridgeport, CT, Internet Archives, https://archive.org/details/Whitebronzemonu00Monu/, last accessed Nov 26, 2020.
3. The Evening Kansas-Republican, Newton, KS Oct 4, 1899.
4. Iowa Civil War Monuments, http://www.iowacivilwarmonuments.com/, last accessed Dec. 1, 2020.

5 The Moline Journal, Moline, Ill, January 19, 1905.

6 The Rock Island Argus, Rock Island, IL, June 17, 1905, 2.

Chapter 6 - Federal Resources

1. https://www.archives.gov/research/military/army/compiled-military-service-records.
2. National Archives, Military Records, https://www.archives.gov/research/military/civil-war/union/veteran-headstones.html, last accessed June 2, 2021.

Chapter 7 - State Records

1. *Essential Civil War Curriculum, Regimental Histories*, Timothy J. Orr, last accessed June 4, 2021, https://www.essentialcivilwarcurriculum.com/regimental-histories.html.
2. *A Short History of the U.S. Army Adjutant General's Corps 1775-2013*, Stephen E. Bower, Ph.D., Staff Historian, U.S. Army Soldier Support Institute, 2013, 15.
3. *Shiloh: War of the Rebellion: Official Records of the Union and Confederate Armies*, U.S. Military Academy, https://usma.libguides.com/Shiloh/Official_Record.
4. Ford, Oren (1933). *A History of the Bounty System Used During the Civil War.* University of the Pacific, Thesis, https://scholarlycommons.pacific.edu/uop_etds/943, last accessed May 17, 2021.
5. *Journal of the Twenty-Fifth National Encampment, Grand Army of the Republic*, (the Tuttle Company, Printers, Rutland, VT) 1891, 77.
6. *Personal War Sketches by Grand Army of the Republic*, Internet Archive, https://archive.org/details/personalwarsketches, last accessed 6/28/2021.
7. Storrs, John W, *The Twentieth Connecticut"; A Regimental History,* (Ansonia, CN, Press of the "Naugatuck Valley Sentinel, 1886) 70-72, https://archive.org/details/twentiethconnect00stor, Last accessed 6/29/2021.

Recommended Reading

Beath, Robert B., *History of the Grand Army of the Republic* (New York, Bryan, Tailor & Co. 1888).

Carnahan, J. Worth , *Manual of the Civil War and Key to the Grand Army of the Republic and Kindred Societies* (Washington D.C., U.S. Army and Navy Historical Association 1899).

Dearing, Mary R., *Veterans in Politics* (Baton Rouge, Louisiana State University Press 1952).

Davies, Wallace Even, *Patriotism on Parade* (Harvard University Press 1955).

McConnell, Stuart, *Glorious Contentment: The Grand Army of the Republic 1865-1900* (University of North Carolina Press 1992).

Jacob, Kathryn Allamong, T*estament to the Union, Civil War Monuments in Washington D.C.* (The Johns Hopkins University Press 1998).

Jordan, Brian Matthew, *Marching Home, Union Veterans and Their Unending Civil Wa*r (New York, Liveright Publishing Corporation 2014).

Gannon, Barbara A., *The Won Cause: Black and White Comrades in the Grand Army of the Republic* (University of North Carolina Press 2011).

INDEX

HERITAGE OF THE GRAND ARMY OF THE REPUBLIC

The Grand Army of the Republic is a unique organization. No child can be born into it. No proclamation of President, edict of King or Czar can command admission. No university or institution of learning can issue a diploma authorizing its holder to entrance. No act of Parliament or Congress secures recognition. The wealth of a Vanderbilt cannot purchase the position. Its doors open only on the presentation of a bit of parchment, worn, torn and begrimed as it may be, which certifies to an honorable discharge from the armies or navies of the Nation during the war against the rebellion, and, unlike any other organization, no new blood can come in. There are no growing ranks from which recruits can be drawn into the Grand Army of the Republic. With the consummation of peace through victory its ranks were closed forever; its lines are steadily growing thinner, and the ceaseless tramp of its columns is with ever lessening tread. The gaps in the picket lines grow wider every day. Details are made for the reserve summoned into the shadowy regions until by and by only a solitary sentinel will stand guard, waiting until the bugle call of the beyond shall muster out the last comrade of the Grand Army of the Republic. (The "solitary sentinel," Albert Woolson, of Duluth, Minn., passed to his reward August 2, 1956.)

Final Journal of the Grand Army of the Republic, 1866 - 1956
United States Government Printing Office, Washington, 1957, Page 1

Books by Gary W. Clark

Genealogical & Historical Research

Cemetery and Gravestone Handbook – Research, Document, and Clean Ancestor Gravestones.

Cemeteries and gravestones are as important to genealogy and family history as census data and vital records, giving us a physical connection to our ancestors. Decipher gravestones to reveal stories. Research veterans in national & local cemeteries. Photograph gravestones as records and art. Clean gravestones to preserve their image.

20th Century Photographs – A Step-by-Step Guide to Identifying and Dating 1900 - 1950 Photos.

20th Century Photographs Photos presents a proven approach to researching and dating 20th century photographs. Decade-by-decade chapters discuss and show how photographs changed over the years, providing you with guidelines to when they were taken. Over 200 full-color beautiful images give you an accurate means to compare your photos.

19th Century Card Photos – A Step-by-Step Guide to Identifying and Dating CDVs and Cabinet Cards.

19th Century Card Photos presents an easy-to-follow set of clues that guides genealogists and historians through the task of dating card photographs from the 1800s. With over 200 high-quality documented photographic examples for every clue, the researcher can accurately and with confidence determine most photograph dates to within two to six years.

Cased Images and Tintypes – A Guide to Dating Daguerreotypes, Ambrotypes, and Tintypes.

The *Cased Images and Tintypes KwikGuide* is a detailed and clear source of identification tips and photo dating information for daguerreotypes, ambrotypes, and tintypes. This is an indispensable reference tool for genealogists, family historians, and photo collectors who are conducting research on vintage 19th century photographs.

Real Photo Postcards – A Guide to Identifying and Dating Real Photo Postcards of the 20th Century.

Real Photo Postcards KwikGuide is a fun guide to the popular postcards in the early 20th century. The KwikGuide presents a short history of all postcards and then covers Real Photo Postcards in detail, including how to recognize and date them. For collectors, determining the date of a photograph may help understand the rarity and value of a postcard.

Photo Restoration – A Step-by-Step Guide for Repairing Photographs with Photoshop Elements.

The *Photo Restoration KwikGuide* presents an easy-to-follow approach to the problem of fixing damage to your valuable photographs. By applying digital editing to high quality scans of the original photographs, anyone can repair the damage by following detailed instructions in the book using the popular Adobe Photoshop Elements software.

Slides and Negatives – Digitize and Protect Your Vintage Film.

Slides and Negatives is a complete guide to digitizing and preserving vintage film for family historians and genealogists. This book examines four primary areas of interest:1) Archiving your slides and negatives for safe keeping, 2) Cleaning slides and negatives, 3) Digitizing the images 4) Identifying and handling potentially hazardous film.

Historical Reflections

Lessons From The Gibson Girl – Her Quest for Equality, Justice, and Love.

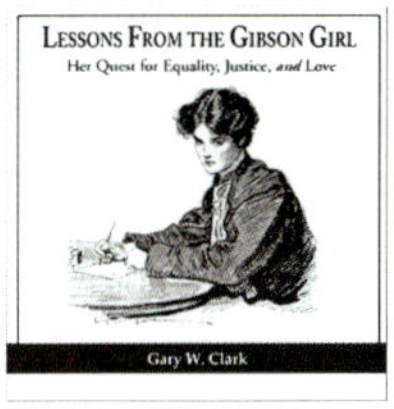

Lessons From The Gibson Girl traces Charles Dana Gibson's fictional heroine as she captured the admiration of men and women with charming yet thought-provoking humor. *Lessons* is a reflective look at his satire and illustrations that lampooned society from the Victorian 1880s through the rowdy 1920s, giving voice to women through the Gibson Girl.

Teenage Soldier – Don K. Clemence, A Greatest Generation Story.

Teenage Soldier shares the story of a young man drafted into the U.S. Army in 1945 while still a high school student. After armored training he was deployed to France where his duties included hunting deserters and Nazi soldiers, and other M.P. duties in postwar Paris – all before turning 20 years old. His story is one of many of the Greatest Generation.

Cruel Irony – Triumphs and Tragedies of a Modern Woman.

The story of a women born on the prairies of Kansas in 1896 who battled family conflicts and tragedies, saw the Great War, devoted her life to the care of others, then experienced her own *Cruel Irony* way too early in life. This is a true story that shows how family history can become an interesting narrative of our ancestors and life around them.

Write a Captivating Family History – Don't Bore Your History to Death, Add Context to the Story.

This book will help you weave historical content into family story. Understanding the weather, the economic and political climate, working conditions, living conditions, and other forces can help expand the story of your ancestors' lives. This book describes the techniques, tips and tricks used in *Cruel Irony* to turn family history into a riveting story.

Additional titles: Fubar2, Archive Photography

Books available from Amazon.com

Visit the author's Amazon page:

www.amazon.com/author/gary-clark

Email: gary@phototree.com

Author's Web: www.garywclark.com

Gary W. Clark with John W. Gallaher's unique gravestone. Not a government supplied stone, but a local production which includes the five-pointed G.A.R. star engraved on the face.

Made in United States
North Haven, CT
05 November 2026

21763956R00077